Cooking as a Chemical Reaction

CULINARY SCIENCE WITH EXPERIMENTS

CRC Press
Taylor & Francis Group
6000 Broken Sound Parkway NW, Suite 300
Boca Raton, FL 33487-2742

© 2015 by Taylor & Francis Group, LLC
CRC Press is an imprint of Taylor & Francis Group, an Informa business

No claim to original U.S. Government works

Printed on acid-free paper
Version Date: 20140502

International Standard Book Number-13: 978-1-4665-5480-1 (Paperback)

This book contains information obtained from authentic and highly regarded sources. Reasonable efforts have been made to publish reliable data and information, but the author and publisher cannot assume responsibility for the validity of all materials or the consequences of their use. The authors and publishers have attempted to trace the copyright holders of all material reproduced in this publication and apologize to copyright holders if permission to publish in this form has not been obtained. If any copyright material has not been acknowledged please write and let us know so we may rectify in any future reprint.

Except as permitted under U.S. Copyright Law, no part of this book may be reprinted, reproduced, transmitted, or utilized in any form by any electronic, mechanical, or other means, now known or hereafter invented, including photocopying, microfilming, and recording, or in any information storage or retrieval system, without written permission from the publishers.

For permission to photocopy or use material electronically from this work, please access www.copyright.com (http://www.copyright.com/) or contact the Copyright Clearance Center, Inc. (CCC), 222 Rosewood Drive, Danvers, MA 01923, 978-750-8400. CCC is a not-for-profit organization that provides licenses and registration for a variety of users. For organizations that have been granted a photocopy license by the CCC, a separate system of payment has been arranged.

Trademark Notice: Product or corporate names may be trademarks or registered trademarks, and are used only for identification and explanation without intent to infringe.

Library of Congress Cataloging-in-Publication Data

Ozilgen, Z. Sibel.
Cooking as a chemical reaction : culinary science with experiments / author, Z. Sibel Ozilgen.
pages cm
Includes bibliographical references and index.
ISBN 978-1-4665-5480-1 (paperback)
1. Food--Composition. 2. Food--Analysis. 3. Chemistry, Technical--Experiments. I. Title.

TX545.O986 2014
664'.07--dc23 2014014149

Visit the Taylor & Francis Web site at
http://www.taylorandfrancis.com

and the CRC Press Web site at
http://www.crcpress.com

Cooking as a Chemical Reaction

CULINARY SCIENCE WITH EXPERIMENTS

Sibel Özilgen

CRC Press is an imprint of the
Taylor & Francis Group, an **informa** business

Contents

PREFACE

This book is written for undergraduate students in culinary arts, nutrition, dietetics, and gastronomy programs. It is intended for students with limited scientific knowledge who are studying different aspects of food preparation and processing. The text uses experiments and experiences from the kitchen rather than theory as the basic means of explaining the scientific facts and principles behind food preparation and processing. Thorough explanations of important scientific concepts that are required to comprehend the text are provided in the glossary.

This textbook is prepared such that students first perform certain experiments and record their observations in tables provided in the book. The science behind their expected observations are then subsequently explained. By conducting experiments and using experiences from the kitchen, this textbook aims to engage students in their own learning process. With this book, students are able to make observations that they will frequently see in the kitchen and will be able to learn the science behind these phenomena. Thus, they will be able to control these phenomena, allowing them to create new food products, improve the quality and safety of their dishes, improve the culinary presentations of their food, and understand what goes wrong in the kitchen.

Many concepts throughout the book are marked with the symbol . This symbol indicates that the concept is an important one that students will come across frequently, both during the study of this text and in the kitchen. The symbol precedes a scientific explanation of the observations made during experiments in the chapter. At the end of each chapter, students are presented with important points to remember, more ideas to try, and study questions to reinforce concepts that were presented in the chapter. It is important to note that it is necessary for students to fully understand the key concepts of each chapter because they will reoccur in subsequent chapters.

Sibel Özilgen

ACKNOWLEDGMENTS

I would like to express my gratitude to the many people who have graciously helped in the preparation of this book

To the Yeditepe University in Istanbul, Turkey, for providing facilities to conduct the experiments.

To Günsel Keserci for the incredible chef graphics.

To my students and colleagues, Dilek Çiftçi, Nurece Yılmazel, Evren Galip Altaylar, Baran Yağmurlu, Aylin Doğan, Merve İşeri, Fikret Soner, and Melek Gündoğdu for conducting experiments, of which photos were taken to be added to the book.

To Semih Özkan, Özcan Yaman, and Barbaros Erdal Çiftçi for food styling and photography.

To Emel Karakaya for her contributions to the figures and photo editing.

To my students for inspiring me to write this book.

To my family for always supporting me.

Finally and, most importantly, to Burak Arda Özilgen for his scientific and stylistic suggestions, comments, and contributions.

About the Author

Sibel Özilgen, Ph.D., is the Head of the Gastronomy and Culinary Arts Department at Yeditepe University, where she has been a faculty member since 2005. Dr. Özilgen completed her Ph.D. degree in food engineering at Middle East Technical University in Turkey. She attended the University of California as a concurrent student during her Ph.D. study. Dr. Özilgen taught classes and conducted research at Massey University in New Zealand and is the author or coauthor of numerous refereed publications and books, one of which concerns the eating habits of preschool children (Alfa Publishing Co., Turkey, 2007). Her research and teaching interests lie in the area of food science, food safety, food product development, and the eating habits of different consumer groups.

CHAPTER 1

Measurements and Units

WHY DO WE NEED MATHEMATICS IN CULINARY PROCESSES?

Measuring the ingredients, calculating the proportions of the ingredients, recipe conversion, resizing a recipe for a number of servings, determination of the cost per meal, and portioning all involve basic math operations, such as addition, subtraction, multiplication, and division.

An inaccurate measurement of the ingredients is one of the major reasons for the success or failure of a recipe. In addition, imprecise measurements and conversion calculations increase food costs. Inaccurate portioning causes consumer dissatisfaction. Therefore, it is important that all chefs must understand basic math operations.

TYPES OF MEASUREMENTS IN THE KITCHEN

There are two types of measurements in culinary operations: qualitative observations and quantitative measurements.

Qualitative observations involve using the five senses to observe and describe the properties of food products (Table 1.1). For example, observing the changes in the color of onions during caramelization, comparing the textures of products

TABLE 1.1
The Most Common Terms That May Be Used in Qualitative Observations

Attribute of the Food Product	The Common Words That May Be Used to Describe the Given Attribute
Appearance	Appetizing, Attractive, Shiny, Cloudy, Greasy, Clear, Opaque, Moist, Dry, Thick, Foamy, Crumbly, Dark, Sticky
Taste	Bitter, Sweet, Sour, Salty, Bland, Spicy, Smoky, Sharp, Tangy, Aromatic, Creamy, Astringent, Nutty
Smell	Floral, Fresh, Smoky, Spicy, Burnt, Strong, Vanilla, Fruity, Nutty
Texture	Gritty, Smooth, Hard, Slimy, Moist, Dry, Spreadable, Chewy, Firm, Fizzy, Foamy, Thick, Thin, Mushy, Spongy, Elastic, Brittle, Crumbly, Crunchy
Sound	Crunchy, Fizzy, Sizzling, Bubbling

prepared with different types of starches, or describing the sweetness of the fruit juices involves qualitative observations. Actual measurements and numbers are *not* involved in qualitative measurements.

Quantitative measurements involve numbers. Instruments, such as kitchen scales, measuring cups, measuring spoons, rulers, and thermometers, are used for quantitative measurements.

Liquids are measured by *volumes*. The most common volume measuring tools in culinary processes are shown in PIC 1.1 and PIC 1.2.

Solids are measured by *weight*. The most common dry weight measuring tools in culinary processes include those shown in PIC 1.3 and PIC 1.4.

Temperature is measured by *thermometers*. The most common thermometers include those shown in PIC 1.5, PIC 1.6, and PIC 1.7.

Pic 1.1
Graduated cups.

Pic 1.2
Measuring cups and spoons.

Pic 1.3
Scales.

Pic 1.4
Measuring cups and spoons.

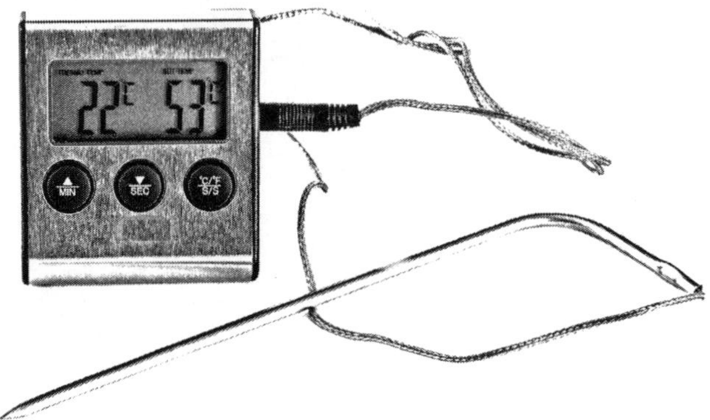

Pic 1.5
Thermometer.

Pic 1.6
Thermometer.

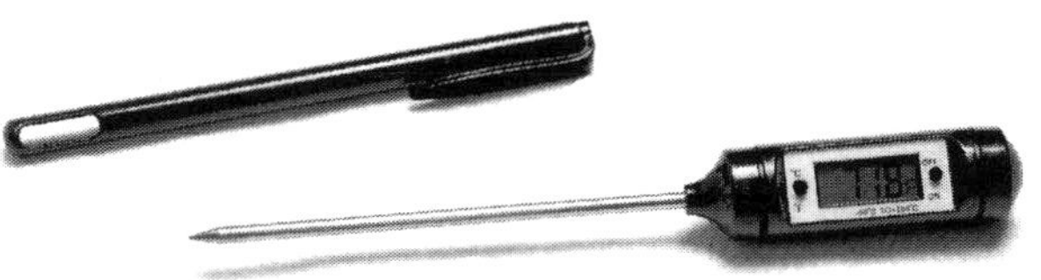

Pic 1.7
Thermometer.

UNITS OF MEASUREMENTS IN CULINARY CALCULATIONS

In every quantitative measurement, there is a number followed by a unit.

Example: A bag of rice is 5 kg (11 lbs).

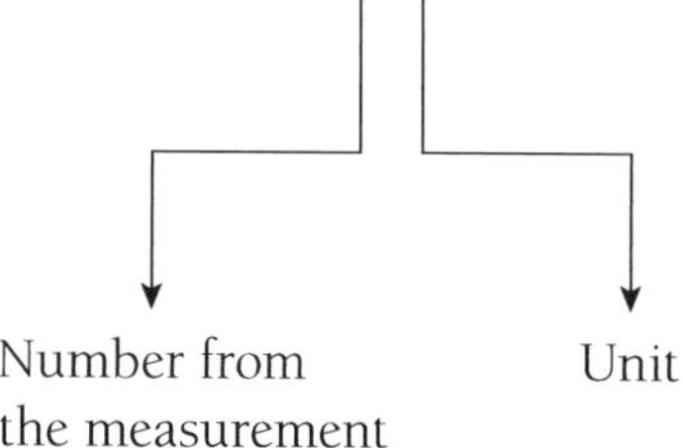

The metric system is the most widely used units of measurements in the world. International System of Units (SI) is based on the metric system. The U.S. Customary System is widely used as the units of measurements in the United States (Table 1.2).

Unit uniformity is important in culinary calculations.

Unit uniformity is important to understand and compare quantities. Accurate measurement of quantities and their units provide:

- standards and stability to the recipes;
- effective inventory and cost control; and
- standard portion size.

Conversion between units of measures is possible.

It is important to know how to convert mass, volume, temperature, and length measured in one unit to the corresponding measure in a different unit. The conversion is an easy mathematical operation because the relationship between different units is constant. Those constants are known as *conversion factors* and they are tabulated in Table 1.3 and Table 1.4.

TABLE 1.2
The Most Common Units of Measures Used in Culinary Processes

Measurement	U.S. Customary System of Units	International System of Units, SI
Length	inch (in)	meter (m)
Volume	fluid ounce (fl. oz); gallon (gal)	liter (l)
Weight	pound (lb), ounce (oz)	gram (g)
Temperature	Fahrenheit, °F	Celsius, °C

TABLE 1.3
The Widely Used Conversion Factors for the Measurements of Length, Volume, and Weight

Measurement	Conversion Factor
Length	1 in = 2.54 cm 1 m = 100 cm
Volume	1 L = 1,000 ml 1 L = 33.8 fl. oz (U.S.) 1 L = 0.264 gal (U.S.)
Weight	1 kg = 1,000 g 1 lb = 454 g 1 lb = 16 oz (U.S.)

TABLE 1.4
The Widely Used Conversion Factor for the Measurements of Temperature

Measurement	Conversion Equation
°C	$T(°F) = 1.8 \times T(°C) + 32$

RULE

To convert one unit to the corresponding measure in a different unit, the number in original units is multiplied by a conversion factor to produce a result in the desired units.

$$\text{Number in original units} \times \left[\textit{conversion factor} \left(\frac{\text{new unit}}{\text{original unit}} \right) \right]$$

$$= \text{New number in new unit}$$

EXAMPLE 1.1

Convert the length of 5.6 inches to its equivalent in units of meters.

Solution

From the equation:

$$5.6\,\cancel{\text{in}} \times \left[2.54 \left(\frac{\cancel{\text{cm}}}{\cancel{\text{in}}} \right) \times \frac{1}{100} \left(\frac{\text{m}}{\cancel{\text{cm}}} \right) \right] = 0.142\,\text{m}$$

EXAMPLE 1.2

Convert the mass of 12 pounds to its equivalent in units of grams.

Solution

From the equation:

$$12\,\cancel{\text{lb}} \times \left[454 \left(\frac{\text{g}}{\cancel{\text{lb}}} \right) \right] = 5448\,\text{g}$$

EXAMPLE 1.3

Convert the oven temperature 303°F in units of °C (Figure 1.1).

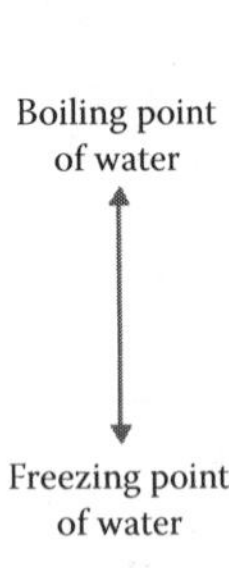

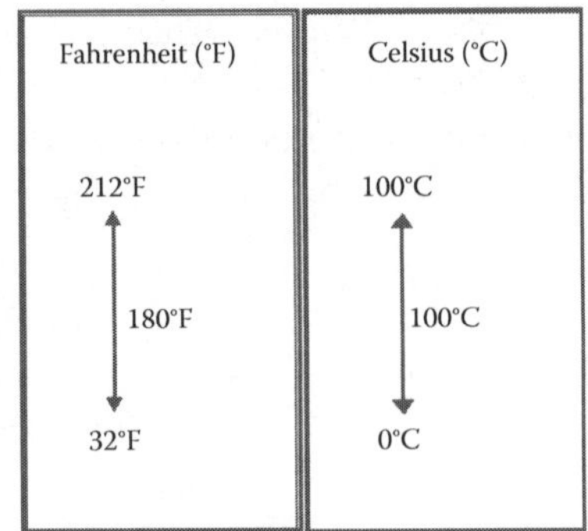

FIGURE 1.1
TEMPERATURE SCALES.

Solution

From the equation given in Table 1.4:

$$T(^\circ C) = \frac{(303^\circ F - 32)}{1.8} = 150.6^\circ C$$

In addition to unit conversion, sometimes it is necessary to convert from volume measures to weight in order to standardize the recipes, especially when adapting a recipe from another source.

For example, although solids are generally measured in weights, they are given in volumes in some recipes, or some measurements are given in weights and some are given in volumes in the same recipe.

The most practical volume-to-weight conversion method in the kitchen can be explained as:

1. Weigh the measuring tool (e.g., cup) and record its weight.
2. Fill the measuring tool with the ingredient as directed by the recipe.
3. Weigh the measuring tool filled with the ingredient.
4. Subtract the first weight from the last measurement.

EXAMPLE 1.4

Calculate the weight of 300 ml ketchup.

Solution

1. Weigh the liquid measuring cup and record it.
2. Fill the measuring cup with 300 ml ketchup.

3. Return the filled measuring cup to the scale and record the weight.
4. Subtract the weight of the liquid measuring cup (from step 1) from the weight of filled measuring cup (from step 3) to find the weight of 300 ml ketchup.

Note: It is possible to convert the measurements given as spoonful, glass, cup, etc., into weight measurements using the same method.

EVERYDAY CALCULATIONS IN THE KITCHEN

FRACTIONS AND PERCENTAGES

Fraction can be defined as the number of parts taken out of a whole quantity divided into equal parts.

It can be calculated as:

$$\text{Fraction} = \frac{\text{number of identical parts being taken from the whole quantity}}{\text{total number of equally divided parts in the whole quantity}}$$

Percentages (%) can be defined as a fraction multiplied by 100.

$$\text{Percentages (\%)} = \text{Fraction} \times 100$$

Fractions are widely used with measurements of ingredients or with portioning, such as 2½ cups of flour or in portioning (⅔ of a pizza). Percentages are used most commonly to calculate food costs and kitchen and restaurant space requirements. Fraction and percentage calculation knowledge also are required in converting and yielding the recipes.

EXAMPLE 1.5

A whole apple is sliced into eight equal pieces. What is the fraction of each slice in the whole apple?

Solution

From the equation:

$$\text{Fraction of each slice} = \frac{1}{8}$$

EXAMPLE 1.6

A chef prepared an apple pie and divided it into 16 equal slices. He sold ¾ of the pie in the morning. Calculate the number of slices sold and the number of slices left.

Solution

Three quarters (¾) means that the whole apple pie is divided into four equal parts and three parts are sold. Because the number of total slices is given, it is possible to calculate the number of slices in each part.

$$\text{Number of slices each part has} = \frac{16\ (\text{slices})}{4\ (\text{parts})} = 4\ \text{slices / part}$$

$$\text{Number of slices sold} = 4\left(\frac{\text{slices}}{\text{part}}\right) \times\ 3\ (\text{parts sold}) = 12\ \text{slices sold}$$

$$\text{Number of slices left} = \text{Total number of slices} - \text{Number of slices sold}$$

$$\text{Number of slices left} = 16\ (\text{slices}) - 12\ (\text{slices}) = 4\ \text{slices left}$$

EXAMPLE 1.7

A chef prepared ice creams in two different flavors. Calculate the number of consumers who preferred to have strawberry-flavored ice cream, if 20% of the 500 consumers had the other flavor.

Solution

From the equation:

$$\text{Number of consumers preferred the other flavor}$$

$$= 500\ (\text{consumers}) \times \frac{20}{100}\left(\frac{\text{consumers preferred the other flavor}}{\text{consumers}}\right)$$

$$= 100\ \text{consumers.}$$

$$\text{Number of consumers preferred the strawberry flavor}$$

$$= 500\ (\text{consumers}) - 100\ (\text{consumers})$$

$$= 400\ \text{consumers}$$

EXAMPLE 1.8

Calculate the percentage of the consumers who preferred to have strawberry-flavored ice cream in the previous example, if 87 of the consumers had the other flavor.

Solution

$$\begin{aligned}&\text{Number of the consumers}\\&\text{preferred strawberry flavor} = 500\ (\text{consumers}) - 87\ (\text{consumers})\\&\qquad = 413\ \text{consumers}\end{aligned}$$

From the equation:

$$\begin{aligned}&\%\ \text{of the consumers}\\&\text{preferred the strawberry flavor} = \frac{413}{500}\left(\frac{\text{consumers}}{\text{consumers}}\right) \times 100\\&\qquad = 82.6\%\end{aligned}$$

YIELD PERCENT

Yield percent can be defined as the percentage of a whole food item that is usable in production of the meal after any required operations (i.e., cutting, peeling) have been completed. Yield percent can be calculated as:

$$\text{Yield percent} = \frac{\text{Edible portion quantity (EPQ)}}{\text{As-purchased quantity (APQ)}} \times 100$$

Method to calculate the yield percent can be explained as:

1. Weigh the ingredients to find the as-purchased quantity (APQ).
2. Weigh the ingredients after cleaning, peeling, etc., to find the edible portion quantity (EPQ).
3. Use the equation to calculate the yield percent.

EXAMPLE 1.9

A recipe requires 5 kg of cleaned, peeled, and diced cucumbers. What is the minimum amount of cucumbers to purchase if the yield percent is 82?

Solution

From the equation:

$$\text{As-purchased quantity (APQ)} = \frac{\text{Edible portion quantity (EPQ)}}{\text{Yield percent}} \times 100$$

$$\text{APQ} = \frac{5\ \text{kg}}{82} \times 100$$

$$= 6.098\ \text{kg cucumber is required}$$

Recipe Yield Conversion

Recipe yield conversion can be defined as adjusting the recipe to increase or decrease the amount that the recipe actually yields.

The amount of each ingredient in the recipe is multiplied by the recipe conversion factor (RCF) to convert the recipe. The RCF can be calculated as:

$$\text{Recipe Conversion Factor (RCF)} = \frac{\text{New yield}}{\text{Original yield}}$$

Example 1.10

The following ingredients list is given for a recipe that yields 50 apple rolls. Convert the recipe to yield 160 rolls.

Ingredients:

- 300 g (10.5 oz) butter
- 30 g (1.05 oz) egg
- 300 g (10.5 oz) yogurt
- 150 g (5.3 oz) corn starch
- 250 g (8.8 oz) sugar
- 1.1 kg (2.43 lb) flour
- 360 g (12.7 oz) apple

Solution

From the equation:

$$\text{RCF} = \frac{160}{50}$$

$$= 3.2$$

TABLE 1.5
Ingredients List for Recipes That Yield 50 Apple Rolls and 160 Apple Rolls, Respectively.

Ingredients	Amounts for 50 Apple Rolls (g/oz)	Amounts for 160 Apple Rolls (g/oz)
Butter	300/10.5	960/33.8
Egg	30/1.05	96/3.4
Yogurt	300/10.5	960/33.8
Corn starch	150/5.3	480/16.9
Sugar	250/8.8	800/28
Flour	1,100/38.8	3,520/124
Apple	360/12.7	1,152/40.6

Multiply all the ingredients by the RCF to calculate the new amounts. If the units of the measurements are not uniform, convert all quantities to the same units.

- Butter: 300 g × 3.2 = 960 g (33.8 oz)
- Egg: 30 g × 3.2 = 96 g (3.9 oz)
- Yogurt: 300 g × 3.2 = 960 g (33.8 oz)
- Corn starch: 150 g × 3.2 = 480 g (16.9 oz)
- Sugar: 250 g × 3.2 = 800 g (28 oz)
- Flour: $(1 \text{ kg} \times \left(\frac{1000 \text{ g}}{1 \text{ kg}}\right) + 100 \text{ g}) \times 3.2 = 3520 \text{ g}$ (7.76 lb) = (18.3 oz)
- Apple: 360 g × 3.2 = 1,152 g (2.5 lbs)

Note: This method is used to convert the amount of *ingredients only*. Do not apply the RCF to convert the processing temperature and time.

SIMPLIFIED STATISTICS FOR CULINARY OPERATIONS

Is it possible to guess how many of each course will be served during weekdays? What is the average number of consumers that will dine tonight? Does the target consumer group like my new food product? What is the average weight of medium sized eggs? Answers are possible with basic statistics knowledge.

Statistics are applied to analyze the quantitative data obtained from the samples of measurements or observations. Culinary professionals use the results of the statistical analysis generally to conclude the observations, to design service operations, to design the menu, to develop new food products, to standardize recipes, and to understand consumer behaviors.

EXPERIMENT 1.1

OBJECTIVE

To explain how to apply basic statistical analysis.

Ingredients and Equipment

- 5 egg cartons with 30 eggs in each
- Kitchen scale

Method

1. Open the first egg carton and count the eggs.
2. Weigh 10 eggs separately and write the measurements in Data Table 1.1.
3. Repeat the same procedure for each egg carton.
4. Calculate the statistical parameters given in Data Table 1.1.

The MEAN can be defined as the average of all scores obtained from the measurements. It can be calculated as:

$$\text{Mean} = \frac{\text{the sum of all scores}}{\text{total number of measurements}}$$

EXAMPLE 1.11

Calculate the mean of observations 2, 3, 4, 6, 4, 5, 3.

Solution

From the equation:

$$\text{Mean} = \frac{2+3+4+6+4+5+3}{7} = 3.86$$

The MODE can be defined as the most frequently occurring score. It is useful when differences between the scores are insignificant.

EXAMPLE 1.12

What is the mode of observations 3, 3, 3, 4, 4, 4, 4, 4, 4, 5, 5, 5, 6?

Solution

The answer is 4, because it is the most repeated number in the observation.

The RANGE can be defined as the difference between the most extreme data values. It is the measure of the spread of the data values. It can be calculated as:

$$\text{Range} = \text{highest point/value} - \text{lowest point/value}$$

Example 1.13

Calculate the range of observations 2, 10, 4, 6, 3.

Solution

From the equation:
Range = 10 − 2 = 8

DATA TABLE 1.1

	Weight of Eggs (g)										Statistical Parameters		
Egg Box	Egg # 1	Egg # 2	Egg # 3	Egg # 4	Egg # 5	Egg # 6	Egg # 7	Egg # 8	Egg # 9	Egg # 10	Mean	Mode	Range
# 1													
# 2													
# 3													
# 4													
# 5													

POINTS TO REMEMBER

- *There are two types of measurements in culinary operations: qualitative observations and quantitative measurements.*
- *Qualitative observations involve the five senses.*
- *Quantitative measurements involve numbers.*
- *In every quantitative measurement, there is a number followed by a unit.*
- *Conversion between units of measures is possible.*
- *Unit uniformity is important in culinary calculations.*
- *Statistics are applied to analyze the quantitative data obtained from the samples of measurements or observation.*

Study Questions

1. Convert the measuring units as indicated:
   a. 10 kg = ___ lb
   b. 160 °F = ____ °C
   c. 8 L = ____ U.S. gal
   d. 4360 ml = ____ L
   e. 678 g = ____ kg
2. You purchased 2,500 g (5.5 lbs) of squash from the market. After cleaning and peeling, you are left with 1,870 g (4 lbs) of squash. Calculate the yield percent for the squash.
3. The shell loss for walnuts is 35%. What is the yield percent?
4. A recipe for tomato soup makes 90 servings. What is the recipe conversion factor if you will be making 200 servings?
5. Calculate the average (mean) weight of a medium-sized apple from the measurements given in Table 1.6.
6. A chef is making meatball sauce and plans on serving two spoonfuls of the sauce with freshly fried meatballs. Each spoonful of sauce weighs 32 grams (1.1 oz). How many kilograms of meatball sauce are needed if the chef plans to serve 600 meatballs?

TABLE 1.6
Weight of Apples.

Sample Number	Weight of the Apple (g/oz)
1	105/3.7
2	102/3.6
3	104/3.66
4	106/3.73
5	101/3.5

SELECTED REFERENCES

Besterfield, D. H. 2009. *Quality control,* 8th ed. London: Pearson Education, Inc.

Blocker, L., and J. Hill. 2007. *Culinary math*, 3rd revised and expanded ed. Hoboken, NJ: John Wiley & Sons.

Boeree, G. Descriptive statistics. Online at: http://webspace.ship.edu/cgboer/descstats.html.

Child Nutritional Programs (CNP) Manager's Manual. *Weighing and measuring*. Online at: http://cnp.alsde.edu/nslp/manuals/CNPManagersManual/Weighing%20and%20Measuring.pdf.

Geankoplis, C. J. 2008. *Transport processes and separation process principles* (includes unit operations), 4th ed. London: Pearson Education, Inc.

Glendale Union High School District Culinary Math Workbook. 2007. Available as e-book: http://aspdf.com/ebook/culinary-math-workbook-pdf.html.

Jones, T. 2008. *Culinary calculations: Simplified math for culinary professionals*, 2nd ed. Hoboken, NJ: John Wiley & Sons.

Office of Mathematics, Science, and Technology Education (MSTE). *Introduction to descriptive statistics*. University of Illinois, Urbana-Champaign. Online at: http://mste.illinois.edu/hill/dstat/dstat.html.

CHAPTER 2

Basic Food Chemistry

FOOD PROCESSING IS ALL ABOUT CHEMISTRY

Foods are made up of chemical compounds that include water, proteins, lipids, carbohydrates, vitamins, and minerals. The composition of the foods determines both physical and chemical properties of the foods. During preparation and processing, such as cooking and drying, a series of chemical reactions occurs in foods. The structures of the chemical compounds are changed and new food products are formed. These types of irreversible changes are called ***chemical changes***.

Basic food chemistry primarily deals with

1. the chemical composition of foods;
2. the chemical structures and the properties of primary food compounds, which are water, proteins, lipids, and carbohydrates; and
3. the chemical changes that the foods undergo during food processing, storage, and transportation.

Knowledge of basic chemistry concepts is required to understand basic food chemistry.

An *atom* is the smallest unit of an element. Atoms are made up of three particles: protons, neutrons, and electrons (Figure 2.1). Protons and neutrons are found

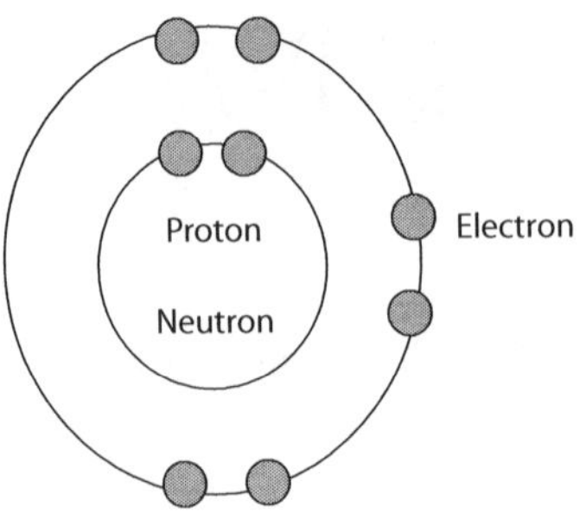

FIGURE 2.1
STRUCTURE OF AN ATOM.

in the nucleus, while the electrons are arranged in shells around the nucleus. The electrons found in the outer shells are called the ***valance*** electrons.

Atoms are connected to each other by ***chemical bonds*** to form ***molecules***, such as NaCl (sodium chloride) and H_2O (water). The valance electrons can be used to form the chemical bonds. There are different types of bonds that hold atoms of the molecules together. Ionic bonds, covalent bonds, and hydrogen bonds are the major bonds that exist in foods.

An ***ionic bond*** is a complete transfer of electrons from the outer shell of one atom to the outer shell of the other atom. In ionic bonds, one atom is negatively charged due to gain of electrons and the other one is positively charged due to loss of electrons. The electronegativity holds the atoms together (Figure 2.2). NaCl is a most common example of the molecule formed by an ionic bond.

A ***covalent bond*** is a complete sharing of one or more electrons between two atoms. It does not involve transfer of electrons (Figure 2.3). One pair of electrons is shared to form one covalent bond. The bonds between the oxygen atom and hydrogen atoms in water molecules are the most common example of a covalent bond.

A ***hydrogen bond*** is a special type of bond in which a hydrogen atom of one molecule is attracted to an electronegative atom. For example, this type of bond exists between water molecules (Figure 2.4).

A ***compound*** is formed if types of atoms in the molecule are different from each other. For example, NaCl is a compound because it has two different atoms in the structure, Na (sodium) and Cl (chlorine). An ***element*** is formed if the types of the atoms in the molecule are the same, such as O_2. Thus, all compounds and elements are molecules.

The structure of the molecules, hence the types of the atoms and the chemical bonds that form the molecules, determine how food will behave during preparation and processing.

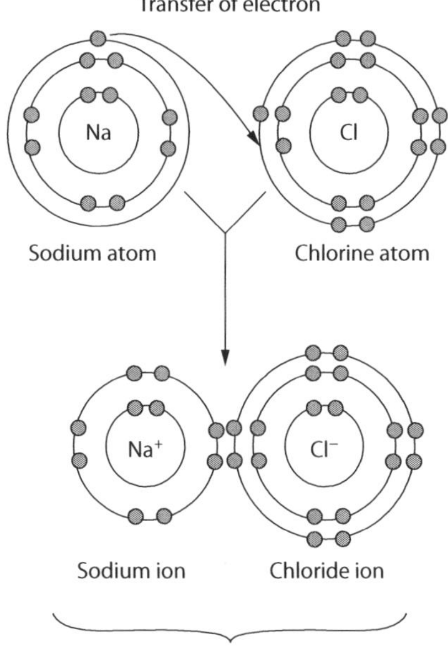

FIGURE 2.2
IONIC BONDING.

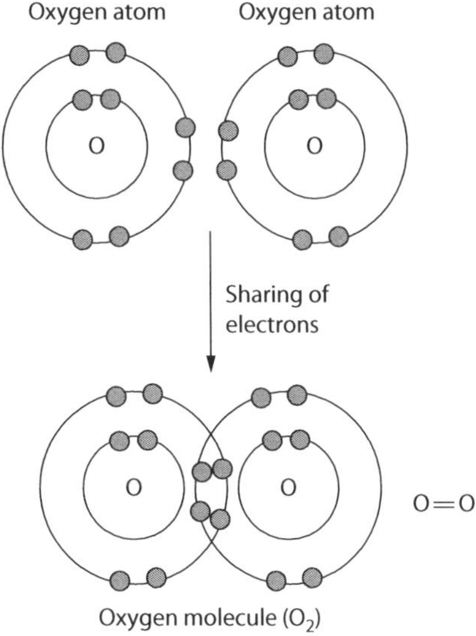

FIGURE 2.3
COVALENT BONDING.

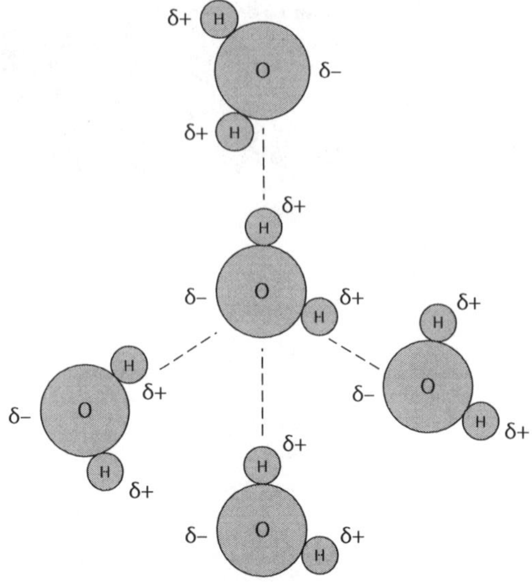

FIGURE 2.4
HYDROGEN BONDING.

EXPERIMENT 2.1

OBJECTIVE

To show food processing is all about chemistry.

Ingredients and Equipment

- 2 cups flour
- 1½ teaspoon baking soda
- 1 cup egg yolks
- ½ teaspoon lemon juice
- Whisk
- Large mixing bowl

Method (PIC 2.1)

1. Sift the flour.
2. In a bowl, beat egg yolks until thick.
3. Then add flour into the egg mixture and fold in.
4. Add baking powder.
5. Record your observations in Data Table 2.1.
6. Add lemon juice into the egg mixture.
7. Record your observations in Data Table 2.1.

Pic 2.1

Reaction between baking powder and lemon juice.

DATA TABLE 2.1

Samples	Observation
Mixture without lemon juice	
Mixture with lemon juice	

THE SCIENCE BEHIND THE RESULTS

Food processing involves a series of chemical reactions.

Each food ingredient in the recipe has its own chemical composition. Therefore, each ingredient serves a purpose in culinary processes. For example, baking soda and baking powder are both leavening agents used in baking. They produce carbon dioxide and cause baked goods to rise. Both require an acid ingredient to give a reaction and to produce carbon dioxide. ***Baking soda*** does not contain an acidic ingredient, therefore, it requires an acidic ingredient in the recipe to produce carbon dioxide. When an acidic ingredient, such as lemon juice, yogurt, or vinegar, is added, a chemical reaction occurs between the baking soda and the acidic ingredient, and carbon dioxide gas is produced. ***Baking powder*** contains baking soda and an acidic ingredient in the same package. When liquid is added, they come together and give a reaction to produce carbon dioxide. Although they are both leavening agents, the two are certainly not interchangeable and they need certain other ingredients to perform.

Food production is applied chemistry, therefore, knowing basic properties of the chemical components of foods will help kitchen professionals to

- better understand their cooking;
- create new food products;
- improve the quality and safety of their dishes;
- improve culinary presentations of their foods; and
- understand what goes wrong in the kitchen.

EXAMPLE 2.1

Following are examples of primary functional properties of food components and additives in baking (PIC 2.2).

Fats

1. Enhance the flavor and mouth feel
2. Develop the texture
3. Shorten the dough
4. Form emulsions
5. Transfer heat
6. Develop the appearance

Carbohydrates

1. Give sweet taste
2. Provide structure and texture

Pic 2.2
Puff pastry with cream filling and strawberries.

3. Lower the freezing point
4. Lower the water activity
5. Fat substitutes
6. Undergo reactions that improve the flavors and colors

Water

1. Good cooking medium
2. Greatly affects the texture and the appearance of the foods
3. Dissolves some flavor compounds
4. Medium for chemical reactions
5. Affects the shelf life

Protein

1. Foam formation
2. Gelation
3. Dough formation
4. Flavor development
5. Viscosity control
6. Water binding
7. Color formation

Food Additives

1. Maintain and improve the nutritional quality
2. Preserve and improve quality and freshness
3. Help in processing or preparation
4. Make food more appealing

POINTS TO REMEMBER

 *Foods are made up of chemical compounds that include water, proteins, lipids, carbohydrates, vitamins, and minerals.*

 *Food production is applied chemistry.*

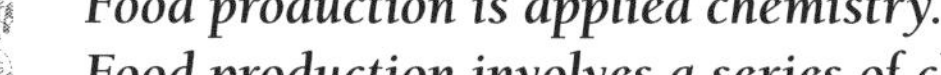 *Food production involves a series of chemical reactions.*

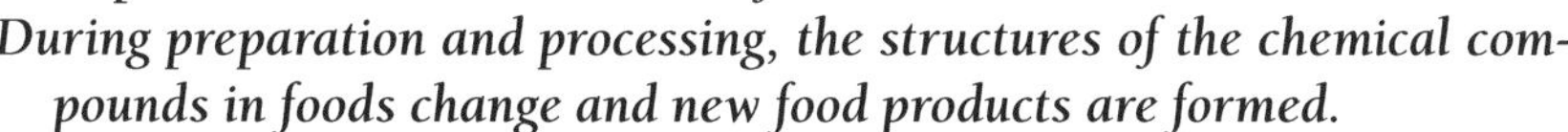 *During preparation and processing, the structures of the chemical compounds in foods change and new food products are formed.*

 *Due to their chemical structures, each food ingredient in the recipe serves a purpose in culinary processes.*

 *Basic knowledge of the chemical components of foods will help kitchen professionals to:*

- better understand their cooking;
- create new food products;
- improve the quality and safety of their dishes;
- improve culinary presentations of their foods; and
- understand what goes wrong in the kitchen.

SELECTED REFERENCES

Elmhurst College, NY. 2003 *Vitual chembook*. What are compound and molecules? Online at: http://www.elmhurst.edu/~chm/vchembook/103Acompounds.html

Fennema, O. R. 1996. *Food chemistry*. Boca Raton, FL: CRC Press.

Gaman, P. M., and K. B. Sherrington. 1996. *The science of food*. 4th ed. London: Elsevier Ltd.

Lasztity, R. 2009. Food quality and standards–Chemistry. In *Encyclopedia of life support systems*. Oxford, U.K.: NESCO Publishing–Eolss Publishers. Online at: http://www.eolss.net/sample-chapters/c10/e5-08-07-00.pdf.

McGee, H. 2007. *On food and cooking: The science and lore of the kitchen*. New York: Scribner.

CHAPTER 3

Water in Culinary Transformations

FUNCTIONAL PROPERTIES OF WATER IN CULINARY PROCESSES

Water is the major constituent of foods. Most natural foods contain 60 to 90% water (Table 3.1).

Food processing, such as drying, cutting, and pressing, decreases the amount of water in foods (Table 3.2).

The primary functions of water in food processing can be listed as:

1. *Water is a good cooking medium*: It conducts heat to the food during cooking. Foods are usually boiled, steamed, braised, or simmered in water.
2. *Water greatly affects the texture and the appearance of foods*: It gives the crisp and moist texture of foods. The appearance of foods can tell consumers what the texture will be when the food is eaten. For example, if the vegetable or fruit, such as lettuce or apple, looks wrinkly, then most probably it has lost its crispy bite.
3. *Water is a good solvent*: It dissolves hydrophilic substances. This property of water primarily enhances the taste of foods because the hydrophilic flavors, such as salt, sugar, and alcohols, dissolve in water. This property of water

TABLE 3.1
Approximate Water Content of Some Natural Foods

Food	Amount of Water in Foods (%)
Tomato	94
Spinach	92
Zucchini	95
Potato	79
Cucumber	96
Broccoli	91
Apple	84
Grape	81
Orange	87
Whole milk	88
Beef	50–70

also enhances the appearance of foods because some color pigments in foods, such as anthocyanins, are also soluble in water.

4. *Water is a medium for chemical reactions*: It hydrates the medium and facilitates the movement of the molecules to give reactions. It also can participate in chemical reactions, such as hydrolysis.
5. *Water content affects the shelf life of foods*: Water is essential for the growth of microorganisms. Most foods contain enough water for the microorganisms to grow. That is primarily why foods with low moisture contents, such as dry beans and some biscuits (cookies), are less perishable than those that have high moisture content.

Understanding the chemical structure of water is crucial for chefs because the functional properties of water are primarily related to its structure.

TABLE 3.2
Approximate Water Content of Some Processed Foods

Food	Amount of Water in Foods (%)
Tomato paste	74
Cereals (rice, wheat, oats)	10–11
Soft cheese, feta cheese	55
Hard cheese, parmesan cheese	29
Bread	34–37
Butter	15
Oil	0

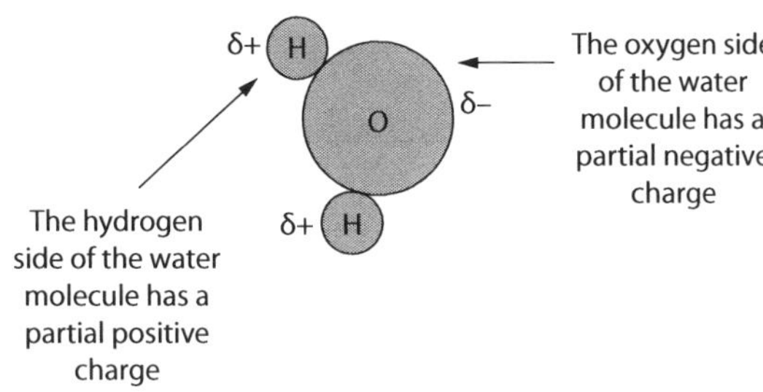

FIGURE 3.1
WATER MOLECULE.

STRUCTURE OF WATER

The water molecule has two hydrogen atoms bonded to one oxygen atom by covalent bonds. The molecular formula is H_2O (Figure 3.1). *It is a polar molecule* where the oxygen atom has a partial negative charge and the hydrogen atoms have a partial positive charge.

Each water molecule can bond with as many as four other water molecules by weak attractions called *hydrogen bonds* (Figure 3.2).

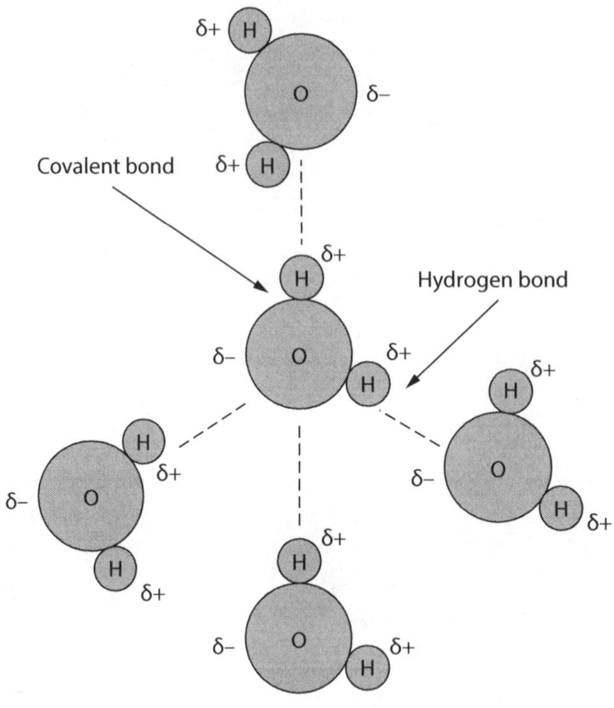

FIGURE 3.2
STRUCTURE OF WATER.

Water is an excellent solvent because it dissolves a variety of different substances. Its polar nature allows water to dissociate ionic compounds into their constituent positive and negative ions. For example, when table salt (NaCl) is mixed with water, the positive part of salt (Na^+) is attracted to the oxygen side of water, while the negative part (Cl^-) is attracted to the hydrogen side; thus salt dissolves in water.

Water is known to exist in three different phases: gas, liquid, or solid (Figure 3.3).

In the ***gas phase (vapor),*** water molecules are well separated and they do not have a regular arrangement. They move freely at high speeds in the gas phase. Water vapor condenses to liquid water as it cools.

FIGURE 3.3
ARRANGEMENT OF WATER MOLECULES IN THE GAS PHASE, LIQUID PHASE, AND SOLID PHASE, RESPECTIVELY.

In the ***liquid phase***, water molecules are closer to each other compared to the gas phase. Once again they do not have a regular arrangement. They move about at lower speeds and slide over each other.

In the ***solid phase***, water molecules are tightly packed in a regular shape. They cannot move from one place to the other.

EXPERIMENT 3.1

OBJECTIVE

To explain the types of water in foods.

Ingredients and Equipment

- 2 apples
- 2 carrots
- 2 trays
- Knife or a slicer
- Cutting board
- Kitchen scale
- Conventional oven
- Baking paper

Method

1. Label two trays "apple tray" and "carrot tray."
2. Line the trays with baking/parchment paper.
3. Measure the weight of each tray and record your results in Data Table 3.1.
4. Peel the apples.
5. Cut each apple into 1 cm × 1 cm × 1 cm (.39 in × .39 in × .39 in) cubes.
6. Arrange the apple cubes on the tray labeled "apple tray" in a single layer.
7. Measure the weight of the apple tray and record your results in Data Table 3.2.
8. Calculate the initial weight of fruits and record your results in Data Table 3.3.
9. Set the oven temperature to 40°C (104°F).
10. Place the tray in an oven.
11. Weigh the tray every 60 minutes and record your results in Data Table 3.2.
12. Repeat steps 9 and 10 until the difference between three consecutive readings is less than 1% of the original fruit mass.
13. Record the time required to dry the apple cubes in Data Table 3.3.
14. Calculate the final weight of apple cubes and record your results in Data Table 3.3.
15. Calculate the total amount of water removed from the apple cubes and record your findings in Data Table 3.3.

DATA TABLE 3.1

Initial Weight of the Apple Tray (g/oz)	Initial Weight of the Carrot Tray (g/oz)

DATA TABLE 3.2

Time (min)	Weight of the Tray with Apples (g/oz)	Weight of the Tray with Carrots (g/oz)
0		
60		
120		
180		
240		
300		
360		
420		
480		

DATA TABLE 3.3

Initial Weight of the Apples (g/oz)	Final Weight of the Apples (g/oz)	Initial Weight of the Carrots (g/oz)	Final Weight of the Carrots (g/oz)	Water Removed from the Apples (%)	Water Removed from the Carrots (%)	Total Time for Drying the Apples (min)	Total Time for Drying the Carrots (min)

Repeat the same procedure with the carrots.

Calculations

1. Initial weight of the foods (g/oz) = initial weight of tray with the foods (g/oz) – weight of the empty tray (g/oz)
2. Final weight of the foods (g/oz) = final weight of tray with the foods (g/oz) – weight of the empty tray (g/oz)
3. Total amount of water removed (g/oz) = initial weight of the foods (g/oz) – final weight of the foods (g/oz)

$$\text{Water removed } (\%) = \frac{\text{total amount of water removed (g)}}{\text{initial weight of the fruit (g)}} \times 100$$

THE SCIENCE BEHIND THE RESULTS

Water in foods primarily exists in two forms:

1. Free water
2. Bound water

Free water is the form of water that is *available* to support biological and chemical reactions in foods. Free water can act as a solvent. It is freezable water. Free water can be easily removed from the food by cutting, pressing, and drying.

Bound water is *not available* for chemical and biological reactions. Bound water is not a good solvent. It is difficult to freeze bound water under normal food processing conditions. Bound water cannot be easily removed from the food by cutting, squeezing, pressing, and drying because the other food constituents, such as proteins, polysaccharides, and fats, hold it.

Although different foods may have the same initial water content, they may have different final water contents after the application of the same drying treatment (as observed in the experiment). This difference primarily arises from the differences in their free water and bound water contents, hence, from the structural differences between the foods (Table 3.3).

Lowering the amount of free water content of foods is one of the basic methods for controlling the spoilage of foods. Drying, freezing, concentration, and addition of hydrophilic substances, such as salts, to bind free water in the food are the most common processes to reduce the amount of free water in food products.

Free water content of foods also is very important to decide the proper cooking technique (wet cooking or dry cooking) and to predict the amount of water to be released from foods during cooking.

TABLE 3.3
Approximate Chemical Compositions of Apple "without Skin" and Carrot

Raw Apple, Measure: 100 g				Raw Carrot, Measure: 100 g		
Nutrient	Units	Value		Nutrient	Units	Value
Water	g	86.67		Water	g	88.29
Protein	g	0.27		Protein	g	0.93
Total lipid	g	0.13		Total lipid	g	0.24
Ash	g	0.17		Ash	g	0.97
Carbohydrate, by difference	g	12.76		Carbohydrate, by difference	g	9.58
Fiber	g	1.3		Fiber	g	2.8
Sugar, total	g	10.10		Sugar, total	g	4.74
Starch	g	0		Starch	g	1.43

EXPERIMENT 3.2

OBJECTIVE

Two cases are presented to understand the difference between water content and water activity.

Case 1

Ingredients and Equipment

- 4 equal-sized (approximately 10 cm × 10 cm × 2 cm/3.9 in × 3.9 in × 0.79 in) feta cheese slices. They must come from the same natural (with no added preservatives), unsalted feta cheese.
- Salt
- Water
- 4 glass storage containers with lids
- 3 bowls
- Spoon
- Food brush
- Kitchen scale

Method

1. Label the containers "control," "5% salt solution," "10% salt solution," and "20% salt solution."
2. Prepare 5%, 10%, and 20% salt solutions in separate bowls.
3. Place a slice of cheese into each glass storage container.
4. Brush the surface of each slice thoroughly with the salt solution according to the label on the respective glasses; start from low concentration if you are using the same brush for all slices.
5. Cover with lids.
6. Store them at room temperature in a safe place.
7. Check the surface of the samples every day at the same time for signs of mold growth.
8. Record your observations in Data Table 3.4.

Note: To prepare the x% of salt solution, add x g of salt in (100–x) g of water. For example, add 10 g (.35 oz) of salt in 90 g (3.1 oz) of water to make a 10% solution of salt in water.

DATA TABLE 3.4

Time (day)	Mold Growth on the Surface of the Control Sample (without Salt)[a]	Mold Growth on the Surface of the Sample Brushed with 5% Salt Solution	Mold Growth on the Surface of the Sample Brushed with 10% Salt Solution	Mold Growth on the Surface of the Sample Brushed with 20% Salt Solution
1				
2				
3				
4				
...				

[a] Use the terms *no growth*, *not significant sign of a mold growth*, and *significant mold growth* to evaluate the samples.

Case 2

Turkish Apricot Dessert Recipe

Ingredients and Equipment

- 500 g (17.6 oz) apricot
- 60 g (2.11 oz) sugar
- Water
- 1 cup halved walnuts
- 8 tablespoon lemon juice
- 4 pots
- 4 plates
- Measuring cup
- Stove

Method

1. Label the plates "control," "10 g (.35 oz) sugar," "20 g (.7 oz) sugar," and "30 g (1.05 oz) sugar, respectively.
2. Clean the apricots.
3. Put 200 ml (6.7 oz) water, X g sugar, 2 tablespoon lemon juice, and 125 g (4.4 oz) apricots into a pot.
4. Place the pot on the stove and bring to a **boil over medium heat.**

5. Put the apricots onto a plate when they get soft.
6. Bring them to room temperature.
7. Place a walnut half in each apricot.
8. Store the plates at room temperature in a safe place.
9. Check the surface of the samples every day at the same time for the signs of mold growth.
10. Record your observations in Data Table 3.5.
11. Continue observation until observing a mold growth on the surfaces of all samples.

For the purpose of the experiment: Carry out stages 2 to 7 with:

1. No sugar (øg): At stage 5, put the apricots on the plate marked "control."
2. 10 g (.35 oz) sugar: At stage 5, put the apricots on the plate marked "10 g (.35 oz) sugar."
3. 20 g (.70 oz) sugar: At stage 5, put the apricots on the plate marked "20 g (.70 oz) sugar."
4. 30 g (1.05 oz) sugar: At stage 5, put the apricots on the plate marked "30 g (1.05 oz) sugar."

DATA TABLE 3.5

Time (day)	Mold Growth on the Surface of the Control Sample (without Sugar)[a]	Mold Growth on the Surface of the Sample with 10 g (.35 oz) Sugar	Mold Growth on the Surface of the Sample with 20 g (.70oz) Sugar	Mold Growth on the Surface of the Sample with 30 g (1.05 oz) Sugar
1				
2				
3				
4				
...				

[a] Use the terms *no growth, not significant sign of a mold growth*, and *significant mold growth* to evaluate the samples.

THE SCIENCE BEHIND THE RESULTS

The total water content of foods does not provide information on the state of water—if it is "free water" or "bound water." *Water activity (a_w)* expresses the availability of water to support chemical and biological reactions in foods. For easier conceptualization, it is possible to say that water activity is the indication of the ***free water*** in a food.

Water activity ranges from 0 (no free water) to 1.0 (pure water). Usually the products that contain a lower percentage of moisture, such as crackers and dried fruits, have lower water activities. On the other hand, some food products with high moisture content can have very little water activity because of their chemical compositions. The water activities of some foods are given in Table 3.4.

Most bacteria require minimum water activity of 0.90 to grow. At water activity below 0.80, most molds cannot be grown. That means foods with lower water activities are more shelf stable.

TABLE 3.4
Approximate Water Content and Water Activity of Some Foods

Food	Water Content (%)	Water Activity
Pure Water	100	1.00
Fresh Meat	65	0.97
Eggs	75	0.97
Bread	35	0.96
Aged Cheddar	36	0.85
Salami	30	0.83
Honey	18	0.75
Dried Fruit	20	0.60 to 0.70
Wheat Flour	12	0.70

Water activity of foods can be adjusted by

1. Physically removing water from foods, such as squeezing, pressing, and drying;
2. Freezing; or
3. Adding substances that bind water, such as sugar and salt.

EXPERIMENT 3.3

OBJECTIVE

To understand the boiling point temperature and boiling point elevation.

Case 1

Determination of boiling point of pure water.

Ingredients and equipment:

- 60 g (2.11 oz) sugar
- Water
- Kitchen scale
- Measuring cup
- 4 saucepans
- Thermometer
- Stove

Methods

1. Measure 500 ml (1 pt) water and place in a saucepan.
2. Place the thermometer in the saucepan. Thermometer should not touch the sides or the bottom of the saucepan.
3. Heat the water.
4. Record your observations in Data Table 3.6 for every 10°C (50°F) increase in temperature.
5. Record the point that temperature stays constant (does not increase) in Data Table 3.7.

Case 2

Determination of boiling point of sugar solutions with different sugar concentrations.

Methods

1. Label 3 saucepans "saucepan 1," "saucepan 2," and "saucepan 3."
2. Place 200 ml (6.7 oz) of water in "saucepan 1."
3. Add 10 g (.35 oz) of sucrose in water. Mix until all crystals are dissolved.
4. Place 200 ml (6.7 oz) of water in "saucepan 2."
5. Add 20 g (.70 oz) of sucrose to the water. Mix until all crystals dissolve.
6. Place 200 ml (6.7 oz) of water in "saucepan 3."
7. Add 30 g (1.05 oz) of sucrose to the water. Mix until all crystals dissolve.

8. Place the thermometer in the "saucepan 1." The thermometer should not touch the sides or the bottom of the saucepan.
9. Heat the solution in saucepan 1.
10. Record your observations in Data Table 3.6.
11. Record the point that the temperature stays constant (does not increase anymore) in Data Table 3.7.
12. Repeat steps 8 to 11 with other sugar solutions in saucepan 2 and saucepan 3.
13. Compare the results.

DATA TABLE 3.6

	Appearance[a]			
Temperature (°C)	Pure Water	Saucepan 1 (Solution with 10 g/ .35 oz of Sucrose)	Saucepan 2 (Solution with 20 g/ .70 oz of Sucrose)	Saucepan 3 (Solution with 30 g/ 1.05 oz of Sucrose)
40				
50				
60				
70				
80				
90				
100				

[a] Use the terms *stagnant, small bubbles, bubbly, boiling* to describe the appearance of the syrups.

DATA TABLE 3.7

Sample	Boiling Point: Point that Temperature Does Not Increase (°C/°F)
Pure water	
Solution prepared with 10 g (.35 oz) of sucrose	
Solution prepared with 20 g (.70 oz) of sucrose	
Solution prepared with 30 g (1.05 oz) of sucrose	

THE SCIENCE BEHIND THE RESULTS

During cooking, as the temperature increases, molecules gain more kinetic energy and they start to move faster. The intermolecular attractions that hold the molecules together break due to increased molecular motion. As a result, molecules escape from the surface of the liquid, and a liquid becomes a vapor. The vapor molecules that have escaped from the liquid phase apply a pressure on the surface of the liquid. That pressure is known as the ***vapor pressure***. As the temperature of the liquid increases, its vapor pressure also increases because molecules gain more kinetic energy.

Boiling occurs when the vapor pressure of the liquid becomes equal to the environmental pressure surrounding the liquid.

The temperature at which a compound changes from a liquid to a vapor (gas) is called the boiling point temperature. Once the liquid starts to boil, the temperature remains constant.

Pure water boils at 100°C (212°F) at standard pressure (1 atmosphere).

At lower environmental pressures, such as on top of a mountain, water boils at temperatures below 100°C (212°F). In contrast, *at higher environmental pressures*, such as the pressure generated in a pressure cooker, water boils at temperatures above 100°C (212°F).

When a nonvolatile solute, such as sugar or salt, is added to the liquid, the vapor pressure is reduced. This means water molecules will need more kinetic energy to escape from the surface of the liquid. Therefore, the solution needs to be heated to a higher temperature to boil. This phenomenon is called *boiling point elevation*. Degree of elevation depends on the concentration of the solutes in the solution. The higher the concentration of the solutes, the higher the boiling point of the solution.

EXPERIMENT 3.4

OBJECTIVE

To understand the freezing point temperature and freezing point depression.

Ingredients and Equipment

- Orange
- Water
- Ice
- Measuring cup
- 3 glass containers; each is approximately 10 cm (3.9 in) in diameter
- Thermometer
- Big bowl
- Knife
- Juice squeezer
- Strainer

Case 1

Determination of the freezing point temperature of pure water.

Method

1. Place 100 ml (3.3 oz) water into a glass container.
2. Place the glass container in a bowl.
3. Fill the bowl with ice. The level of the ice should be higher than the level of water in the glass.
4. Place the thermometer in the water. Thermometer should not touch the sides or the bottom of the glass container.
5. **During the experiment, constantly stir the water gently with the thermometer to avoid surface ice crystallization.**
6. Measure the temperature every two minutes.
7. Record your readings in Data Table 3.8.
8. Record the measurements at the point that the temperature stays constant (does not decrease) in Data Table 3.9.

Case 2

Determination of the freezing point of orange juice with pulp.

Method

1. Squeeze 100 ml (3.3 oz) of orange juice and place in a clean glass.
2. Place the glass in a bowl.

3. Fill the bowl with ice. The level of the ice should be higher than the level of juice in the glass.
4. Place the thermometer in the juice. The thermometer should not touch the sides or the bottom of the glass.
5. **During the experiment, constantly stir the juice** gently with the thermometer to avoid surface ice crystallization.
6. Measure the temperature every two minute.
7. Record your readings in Data Table 3.8.
8. Record the measurements at the point that the temperature stays constant (does not decrease) in Data Table 3.9.

Case 3

Determination of the freezing point of orange juice with no pulp.

Method

1. Squeeze the oranges.
2. Drain and separate the pulp from the juice.
3. Place 100 ml (3.3 oz) of orange juice in a clean glass.
4. Place the glass in a bowl.
5. Fill the bowl with ice. The level of the ice should be higher than the level of juice in the glass.
6. Place the thermometer in the juice. The thermometer should not touch the sides or the bottom of the glass.
7. During the experiment, constantly stir the juice gently with the thermometer to avoid surface ice crystallization.
8. Measure the temperature every two minutes.
9. Record your readings in Data Table 3.8.
10. Record the measurements at the point that the temperature stays constant (does not decrease) in Data Table 3.9.
11. Compare your results.

Note: Add more ice if it melts during the experiment.

DATA TABLE 3.8

Time (min)	Temperature (°C/°F)		
	Water	Orange Juice with Pulp	Orange Juice with no Pulp
0			
2			
4			
6			
8			
10			
12			
14			
16			
18			
...			

DATA TABLE 3.9

Samples	Time for Freezing (min)	Freezing Temperature: Point that Temperature Does Not Decrease (°C/°F)
Water		
Orange juice with pulp		
Orange juice with no pulp		

THE SCIENCE BEHIND THE RESULTS

As temperature decreases, kinetic energy of the molecules decreases. Therefore, the motion of the molecules slows down and they can get closer. Intermolecular forces let the molecules get into more ordered structure. At a specific temperature, liquids turn into solid. Once the liquid starts to solidify, the temperature remains constant.

Freezing point is the temperature at which a liquid changes to a solid. Pure water freezes at 0°C (32°F).

When a solute, such as salt or sugar, is added to water, the solute physically interrupts the intermolecular forces in the solvent. A solution freezes at a lower temperature than the pure solvent. This phenomenon is called *freezing point depression*. Foods freeze below 0°C (32°F) because they naturally contain solutes, such as fibers, fats, vitamins, and sugar.

EXPERIMENT 3.5

OBJECTIVE

To understand how freezing rate affects the ice crystal size.

Ingredients and Equipment

- 1 kg (2.2 lbs) spinach
- Water
- Ice
- Pot
- Bowl
- Colander
- Blast chiller
- Freezer
- 2 large Ziploc® bags
- Stove
- Kitchen scale

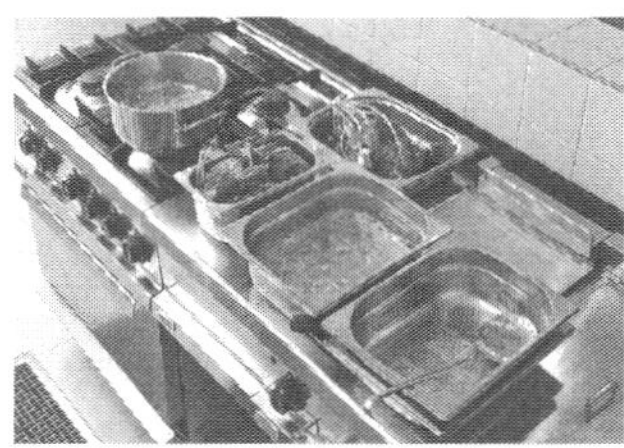

Exp 3.1

Method

1. Label the Ziploc bags "conventional freezer" and "blast freezer."
2. Clean and wash spinach.
3. Boil a large pot of water.
4. Place the spinach into boiling water.
5. Blanch the spinach for two minutes.
6. Place the blanched spinach into a bowl of ice water immediately.
7. Drain the blanched spinach in a colander.
8. Analyze the texture and record in Data Table 3.10.
9. Weigh and equally divide the blanched vegetables into two bags, and close them.
10. Place the first bag in a conventional freezer and leave for freezing. (This stage may take a few hours.)
11. Place the second bag in a blast freezer, sharp freeze, and move to the conventional freezer after full freezing for frozen storage. (Freezing stage takes minutes.)

Exp 3.2

Exp 3.3

Exp 3.4

12. Next day, evaluate the samples in their frozen states.
13. Record your observations in Data Table 3.10.
14. Thaw both samples completely on the counter.
15. Evaluate the samples.
16. Record your observations in Data Table 3.10.
17. Compare the results (EXP 3.1–EXP 3.6).

Exp 3.5

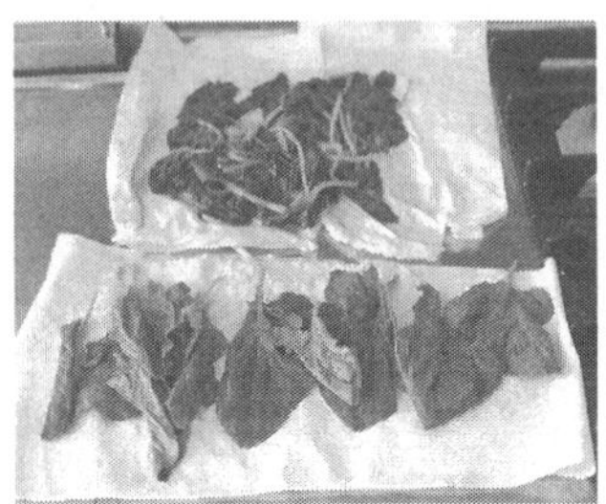

Exp 3.6

DATA TABLE 3.10

	Spinach Frozen in Conventional Freezer	Spinach Frozen in Blast Freezer
Amount of ice crystals on the surface[a]		
Size of ice crystals on the surface[b]		
Texture: Before freezing[c]		
Texture: Frozen		
Texture: After thawing		

[a] Use the terms *insignificant* and *numerous* to evaluate the amount of ice crystals.
[b] Use the terms *tiny, small,* and *coarse* to evaluate the size of the crystals.
[c] Use the terms *crispy, soft*, and *wilted* to evaluate the texture of the samples.

The Science behind the Results

When water freezes its volume expands because the water molecules are locked into an ordered crystalline structure.

As discussed previously, in the solid phase, water molecules are tightly packed in a regular shape. That structure keeps the molecules rigidly apart from one another by hydrogen bonds, which causes large gaps between the molecules. In other words, volume created by the same number of molecules is larger in the solid phase (Figure 3.3) compared with the liquid phase. It is observed as an expansion in the volume of the foods when they are frozen.

This fact is important because:

1. Containers for food freezing must be designed to accommodate the volume increase.
2. The ice crystals may damage the physical structure of foods because of cell rupture, which may occur depending on the ice crystal size and location.

The size and amount of the ice crystals in foods can contribute to several sensory attributes, such as texture and mouthfeel. Large ice crystals in foods may rupture the food cells and may cause a loss of natural juice found in cells upon defrosting. The foods get dry and wilted. Large ice crystals also may give a very gritty and lumpy structure to some frozen foods, such as ice creams and frozen desserts.

Rate of freezing determines the size and the number of ice crystals in frozen foods.

Generally, a slow freezing rate results in a small number of ice crystals. These ice crystals are usually large in size. Sharp (rapid) freezing results in a large number of ice crystals. These ice crystals are usually small in size. Therefore, rapid freezing is usually favored for the food freezing process.

Density can be defined as mass per unit volume of water. Therefore, because ice has fewer molecules than liquid water in an equal volume, it is less dense compared to liquid water. This is why ice cubes floats in water.

EXPERIMENT 3.6

OBJECTIVE

To understand the effects of temperature fluctuation and mixing on the ice crystal size.

Basic Ice Cream Recipe

Ingredients and Equipment

- 1 cup heavy cream
- 3 cups half-and-half cream
- 8 egg yolks
- 1 cup white sugar
- 1/8 teaspoon salt
- Heavy saucepan
- 3 shallow containers
- 2 large bowls
- Ice cream maker
- Stretch film/plastic wrap (to cover the containers)
- Stove
- Whisk
- Refrigerator
- Freezer

Case 1

Effect of temperature fluctuation.

1. Take three shallow containers and label them "sample 1," "sample 2," and "sample 3".
2. Pour the heavy cream and half-and-half into a heavy saucepan.
3. Simmer over medium-low heat, stirring frequently.
4. Turn the heat down to low.
5. Whisk together the egg yolks, sugar, and salt in a large bowl.
6. Slowly pour two cups of hot cream mixture into the egg yolk mixture while whisking constantly and thoroughly.
7. Pour the egg yolk mixture back into the heavy saucepan with the remaining hot cream.
8. Whisk constantly over medium-low heat until the mixture thickens; this stage takes five to eight minutes. Do not boil the mixture.
9. Pour the ice cream mixture into a clean bowl and allow to cool at room temperature for about 20 minutes.

DATA TABLE 3.11

	Response Number										
Sample Number	1	2	3	4	5	6	7	8	9	10	Average Score of Responses
1											
2											
3											

10. Chill overnight in the refrigerator.
11. Next day, place one third of the mixture into the shallow container labeled "sample 3" and cover. Save it for Case 2.
12. Pour the rest of the mixture into an *ice cream maker* that agitates the mixture continuously while freezing it.
13. Freeze both samples thoroughly.
14. After freezing, immediately divide and place the ice cream into two different shallow containers labeled "sample 1" and "sample 2" and cover them.
15. Place both containers in the freezer (–18°C/0°F).
16. After an hour, place "sample 1" in the refrigerator at 5 to 8°C (41–46°F) (melting).
17. After an hour, place "sample 1" back in the freezer and freeze overnight (refreezing).
18. Next day, carry out a tasting session "with sample 1 and sample 2".

    Tasting technique: Ask 10 people to taste a small amount of each ice cream sample and evaluate the texture in a ranking of 1 to 5, where 1 = coarse/icy, 5 = smooth. Collect their responses and prepare Data Table 3.11.
19. Calculate the average scores of the responses.
20. Record the results in Data Table 3.11.

Case 2

Effect of mixing.

1. After stage 11, place the container labeled *sample* 3 in the freezer (–18°C/0°F).
2. Freeze overnight without mixing.

3. Next day, carry out a tasting session as explained in Case 1; record the results in Data Table 3.11.
4. Calculate the average scores of the responses.
5. Record the results in Data Table 3.11.
6. Compare the results of the samples.

Hint:

$$\text{Average score} = \frac{\text{response 1} + \text{response 2} + \text{response 3} + \ldots + \text{response 10}}{10}$$

The Science behind the Results

The smooth and creamy texture is one of the major quality indicators of frozen desserts and it is primarily affected by the *size distribution of the ice crystals* in the product. The texture becomes gritty and rough as the number of larger crystals in the frozen dessert increases.

Rapid freezing and proper agitation (mixing) of foods during freezing are the major requirements for formation of small ice crystals in frozen desserts.

The effects of temperature and the rate of cooling on the number and size of the ice crystals have been discussed previously. *Agitation* physically reduces the size of the crystals. It creates more dispersed ice nucleation because agitation keeps the ice crystals and also water molecules apart from each other. Because they cannot come in contact with each other, they cannot form larger ice crystals. Continuous agitation is required during the processing of the frozen foods to require a smooth texture, such as in ice cream.

Quality loss may occur if the frozen food is not stored properly.

Ice crystals have a tendency to come together to form larger crystals over time during storage. Therefore, *extended storage time* may decrease the quality of frozen foods. *Temperature fluctuations* also enhance the formation of larger crystals. Ice crystals easily melt if the temperature fluctuates and goes above the freezing point during storage. When the temperature drops to the freezing temperature again, the larger ice crystals will be formed as a result of uncontrolled freezing.

EXPERIMENT 3.7

OBJECTIVE

To explain osmosis.

Case 1

Ingredients and Equipment

- 1 large head of lettuce
- Two bowls
- Salt
- Knife
- Cutting board
- Kitchen scale
- 2 colanders

Method

1. Label the bowls "fresh" and "salted."
2. Wash and clean the lettuce.
3. Finely chop the lettuce.
4. Weigh and equally divide the lettuce in to two bowls.
5. Record the initial weight of the individual lettuce in Data Table 3.12.
6. Sprinkle salt over the lettuce in the bowl labeled "salted."
7. Place both bowls in the refrigerator.
8. Every 30 minutes, observe the textural changes in each sample and record your observations in Data Table 3.12.
9. Drain both samples in separate colanders after two hours.
10. Weigh the samples and record the final weights in Data Table 3.12.
11. Calculate the amount of water released from the lettuce samples.

Total amount of water released (g) = initial weight of the lettuce (g) − final weight of the lettuce (g)

Case 2

Strawberry jam

Ingredients and Equipment

- 1 kg (2.2 oz) strawberries
- 1 kg (2.2 oz) sugar
- 2 tablespoon lemon juice
- Container
- Plate

DATA TABLE 3.12

Texture[a]		
	Fresh Lettuce	Salted Lettuce
Time (min)		
30		
60		
90		
120		
Initial Weight (g)		
Final Weight (g)		
Total Amount of Water Released (g)		

[a] Use the terms *crispy* and *wilted* to define the texture of the samples.

- Pot
- Stove
- Stretch film/plastic wrap

Method

1. Wash the strawberries and remove stems and leaves.
2. Place the strawberries in a container.
3. Sprinkle sugar over the fruit and cover the container.
4. Let it stand for six to eight hours.
5. Every hour, observe the release of juice from the fruits.
6. Record your observations in Data Table 3.13.
7. Place the strawberries and the juice into a pot.
8. Place the pot over low heat and cook until you see bubbles.
9. Then turn the heat to medium and simmer for 15 to 20 minutes.
10. Drop some jam on a plate. If the drop keeps its shape when it cools, add the lemon juice, and stir.
11. Cool and store (EXP 3.7–3.10).

DATA TABLE 3.13

Time (min)	Observations[a]
0	
60	
120	
180	
240	
300	
360	

[a] Use your own terms to describe the amount of juice released from the fruit.

Exp 3.7

Exp 3.8

Exp 3.9

Exp 3.10

The Science behind the Results

Water naturally moves from a region of lower solute concentration (higher solvent concentration) to a region of higher solute concentration (lower solvent concentration), through a semipermeable membrane. This phenomenon is called *osmosis*.

Osmosis may decrease the quality of foods. For example, salad greens become wilted after they have been in contact with salt for a while. Similarly, fruits release their juice when they are sprinkled with sugar. The reason is the movement of water in foods through the cell walls to a region of higher solute (salt or sugar) concentration.

On the other hand, osmosis is commonly used to preserve foods. For example, salting (curing) draws moisture out of the foods through osmosis, which *decreases the water activity* of the foods and makes them less susceptible to biological and chemical reactions.

POINTS TO REMEMBER

- *The primary functions of water in food processing include:*
  1. Water is a good cooking medium.
  2. Water greatly affects the texture and the appearance of foods.
  3. Water is a good solvent.
  4. Water is a medium for chemical reactions.
  5. Water content affects the shelf life of foods.
- *Understanding the structure of water is crucial for chefs because the functional properties of water are primarily related to its polar structure.*
- *Polar nature allows water to dissociate ionic compounds into their positive and negative ions.*
- *Water is known to exist in three different phases: solid, liquid, or gas.*
- *Water in foods exists in two forms as:*
  1. Free water
  2. Bound water
- *Availability of water in foods to support chemical and biological reactions is expressed by water activity.*
- *The temperature at which a compound changes from a liquid to a vapor is called the* boiling point.
- *Nonvolatile solutes in the liquid increase the boiling point temperature.*
- *Freezing point is the temperature at which a liquid changes to a solid.*
- *Solutes in the liquid decrease the freezing point temperature.*
- *As water freezes, its volume expands.*
- *Rapid freezing and agitation are the major requirements for the formation of small crystals in frozen desserts.*
- *Extended storage time and temperature fluctuations during storage decrease the quality of frozen food.*
- *Water naturally moves from a region of lower solute concentration to a region of higher solute concentration through cell walls of the foods. This phenomenon is called* osmosis.

More Ideas to Try

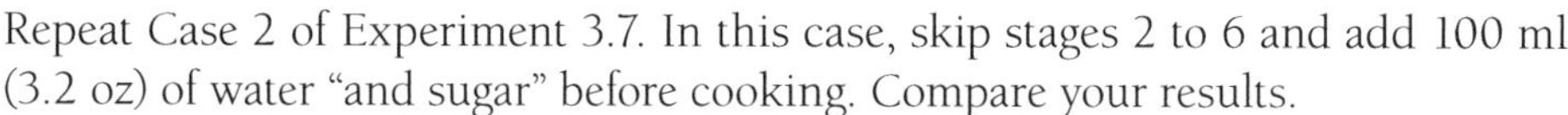

Repeat Case 2 of Experiment 3.7. In this case, skip stages 2 to 6 and add 100 ml (3.2 oz) of water "and sugar" before cooking. Compare your results.

Study Questions

1. What is the importance of water in food preparation?
2. Which one boils first: pure water or salty water?
3. Why do frozen meats, fruits, and vegetables release water upon thawing? How can you minimize it?
4. What makes jam and honey less perishable? Explain.
5. Which one is more perishable: watermelon or beef? Explain.
6. Why are dried food products more stable than fresh ones?
7. What is the reason for the workers at the grocery store spraying water on the vegetables? Explain.

SELECTED REFERENCES

Bastin, S. 1997. *Water content of fruits and vegetables*. Online at: http://www2.ca.uky.edu/enri/pubs/enri129.pdf.

Coultate, T. P. 1996. *Food: The chemistry of its components,* 3rd ed. London/Cambridge: Royal Society of Chemistry (RSC).

Cybulska, E. B., and P. E. Doe. 2006. Water and food quality. In *Chemical and functional properties of food components*, 3rd ed., ed. A. E. Sikorski. Boca Raton, FL: CRC Press.

Gaman, P. M., and K. B. Sherrington. 1996. *The science of food,* 4th ed. London: Elsevier.

McGee, H. 2004. *On food and cooking,* 1st rev. ed. New York: Scribner.

Mudambi, S. R., S. M. Rao, and M. V. Rajagopal. 2006. *Food science,* rev. 2nd ed. New Delhi: New Age Int. Ltd, Publishers.

Sun, D-W. 2011. *Handbook of frozen food processing and packaging*, 2nd ed. Boca Raton: CRC Press.

USDA Nutrient Data Laboratory, Beltsville, MD. Online at: http://fnic.nal.usda.gov/nal_display/index.php?info_center=4&tax_level=2&tax_subject=279&topic_id=1387.

Vaclavik, V. A., and E. W. Christian. 2008. *Essentials of food science,* 3rd ed. Berlin: Springer.

Water. Online at: http://class.fst.ohio-state.edu/fst605/605%20pdf/Water.pdf.

CHAPTER 4

Carbohydrates in Culinary Transformations

FUNCTIONAL PROPERTIES OF CARBOHYDRATES IN CULINARY PROCESSES

Carbohydrates are the organic compounds naturally found in many foods, primarily in plants. They have major roles in food preparation.

The primary functional properties of carbohydrates in culinary processes are listed below.

1. They are the major source of energy.
2. They give foods their sweet taste.
3. They provide structure and texture in food products.
4. They lower the freezing point of food products.
5. They lower the water activity of food products.
6. They are the foods for some microorganisms in fermentation, such as yogurt production.
7. They are used as fat substitutes.
8. They undergo reactions that improve the flavors and colors of certain food products:
   a. Maillard browning
   b. Caramelization

The chemical structure of carbohydrates in foods, molecular conformation, and the number of monomer units in the structure will influence the way the food behaves under different production, preparation, processing, and storage conditions. For example, cooking times, cooking methods, and the final textural properties of rice dishes differ depending on the type of rice used because different types of rice have different carbohydrate compositions. Therefore, the appropriate rice type should be chosen depending on the type of dish that is to be prepared. Similarly, the major difference between potato types is the amount and the nature of the starch each contains. Thus, the potato type chosen for baking is different from the one chosen for boiling.

Understanding the basic structure of carbohydrates is crucial for chefs because the functional properties of carbohydrates are primarily related to their structures.

CARBOHYDRATE STRUCTURE

Carbohydrates are made of carbon (C), hydrogen (H), and oxygen (O).

$$C_n(H_2O)_n$$

C_n = carbo
$(H_2O)_n$ = hydrate

GENERAL FORMULA FOR CARBOHYDRATES

Carbohydrates are classified into three major groups according to the number of simple sugar (monomer) units that they have in their structures.

1. MONOSACCHARIDES ($C_6H_{12}O_6$ OR $C_6(H_2O)_6$)

Monosaccharides contain a single (mono) carbohydrate molecule (Table 4.1). The most common monosaccharides include:

- Glucose (G)
- Fructose (F)
- Galactose (GA)

Glucose (G) is the most common monosaccharide found in foods. It is the primary monomer of polysaccharides. *Fructose (F)* is the sweetest of all the sugars and is

TABLE 4.1
Structures of Monosaccharides and Disaccharides

Monosaccharides	Disaccharides
GLUCOSE 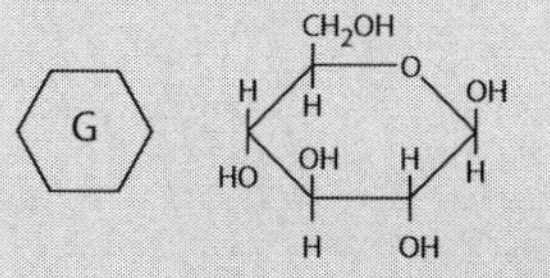 (G)	SUCROSE (G–F) 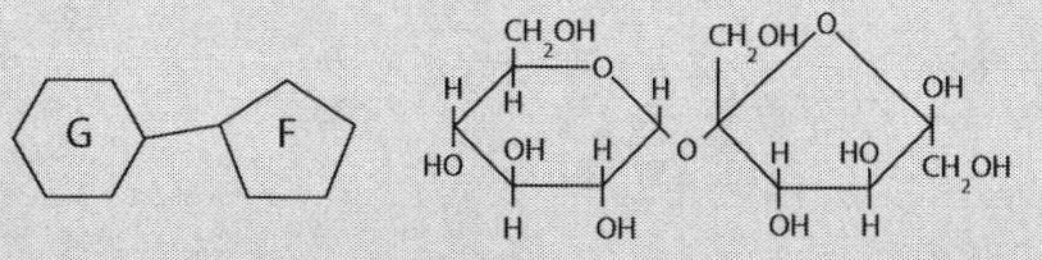
FRUCTOSE (F) 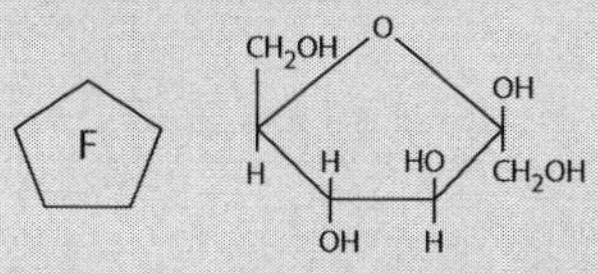	MALTOSE (G–G) 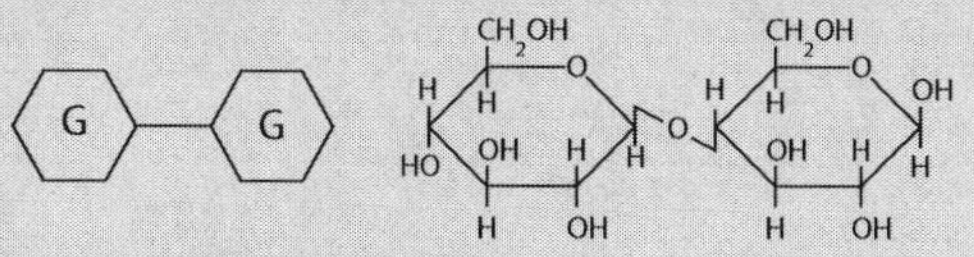
GALACTOSE (GA) 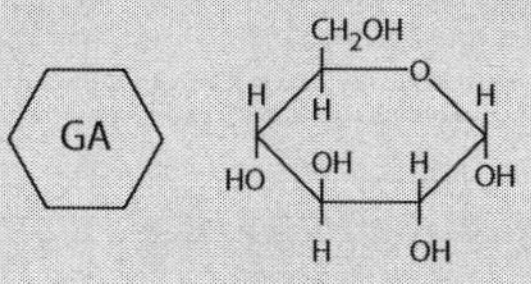	LACTOSE (GA–G) 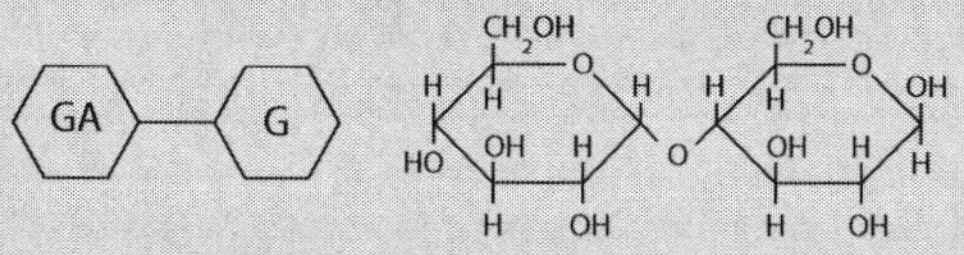

known as "fruit sugar." *Galactose (GA)* is one of the simple sugars found in the structure of milk sugar.

2. Disaccharides ($C_{12}H_{24}O_{12}$ or $C_{12}(H_2O)_{12}$)

Disaccharides are formed when two (di) monosaccharide molecules are chemically linked (bound) together (Table 4.1). The most common disaccharides include:

- Sucrose (G-F)
- Lactose (G-GA)
- Maltose (G-G)

Sucrose is made when one molecule of glucose and one molecule of fructose link together. It is obtained from sugarcane and sugar beet. It is known as "table sugar." *Lactose* is composed of one molecule of glucose and one molecule of galactose. It is found in milk and is known as "milk sugar." *Maltose* is a disaccharide composed of two molecules of glucose. It is found in malted cereals and sprouted grains.

3. Polysaccharides

Polysaccharides are the polymers of the simple sugars. They have more than 10 monomer (monosaccharide) units in their structures. The most abundant polysaccharides in foods can be listed as:

- Starch
- Dietery fibers
- Glycogen (animals)

EXPERIMENT 4.1

OBJECTIVES

- To explain the properties of saturated and unsaturated solutions.
- To determine the saturation points of sugar solutions at different temperatures.

Ingredients and Equipment

- 250 g (8.8 oz) table sugar (sucrose)
- Water at room temperature (25°C/77°F)
- Food coloring (optional)
- Aroma (optional)
- Pots
- Jar
- Measuring cup
- Food thermometer
- Spoons
- Pipe cleaner
- Kitchen scale
- Stove

Method

1. Pour 100 ml (3.3 oz) water into a pot.
2. Measure the temperature of the water (it should be around 25°C/77°F).
3. Add a ***small*** amount of the sugar to the water and stir with a spoon until dissolved.
4. Continue adding the sugar in small amounts until the solute will no longer dissolve. Always completely dissolve the sugar before adding more.
5. Weigh the amount of sugar remaining to determine how much sugar was added to the solution.
6. Calculate the weight of sugar added per 100 ml (3.3 oz) water and record the result in Data Table 4.1.
7. Add a small amount of the sugar to the solution; stir with a spoon.
8. Observe the undissolved sugar crystals at the bottom of the pot (this is called a supersaturated solution).
9. Move the pan to the stove over low heat.
10. Bring the solution to 55°C (131°F) and hold at this temperature. Stir until the sugar dissolves.
11. Repeat steps 3 to 8.

DATA TABLE 4.1

Temperature of the Solution (°C/°F)	Total Weight of Sugar Added per 100 ml (3.3 oz) of Water (g)
25/77	
55/131	
65/149	
75/167	
85/185	
Boiled	

DATA TABLE 4.2

Initial weight of the jar (g/oz)	
Final weight of the jar (g/oz)	
Amount of water evaporated (g/oz)	

Complete the same procedure for solutions at 65°C (149°F), 75°C (167°F), 85°C (185°F), and boiled solution (from stage 10).

12. Pour the final (boiled) solution into a jar while it is hot.
13. Add some food colorings and an aroma of your choice.
14. Tie the pipe cleaner to the pencil.
15. Place the pencil across the neck of the jar, and wrap the pipe cleaner around the pencil until the pipe cleaner is hanging about 2 cm (¾ in) from the bottom of the jar.
16. Weigh the jar and record in Data Table 4.2.
17. Cover the jar with paper towel (do not close the jar).
18. Keep it undisturbed at room temperature for crystals to develop on the pipe cleaner.
19. Record your observations every 24 hours in Data Table 4.3.
20. When the sugar crystal is fully formed, weigh the jar.
21. Calculate the amount of water removed and record in Data Table 4.2.

DATA TABLE 4.3

Time (h)	Crystal Development on the Surface and Bottom of Jar[a]	Crystal Development around the Pipe Cleaner
24		
48		
72		
96		
...		

[a] Observe the amount of crystals forming on the side of the jar. Explain the size, shape, and number of crystals on the string.

Total weight of sugar added per 100 ml (3.3 oz) water = initial weight of sugar (g/oz) – weight of sugar remaining (g/oz)

Amount of water evaporated (g/oz) = initial weight of the jar (g/oz) – final weight of the jar (g/oz)

THE SCIENCE BEHIND THE RESULTS

Monosaccharides and disaccharides are soluble in polar solvents, such as water. Monosaccharides contain polar hydroxyl groups (OH), which can easily form hydrogen bonds with water.

The maximum amount of solute that can be dissolved in a fixed volume of solvent is definite. Any excess solute added will stay as crystals at the bottom of the container. The concentration of solute where no more solute can be dissolved in the solvent is called the ***saturation point*** and the solution is called the ***saturated*** solution, because it cannot accommodate more solute.

The solubility of solids in liquids usually increases with temperature. As the temperature increases, the molecules gain more energy and they move from one position to another more easily. Sugar molecules move apart as a result of increased molecular motion. Increased molecular motion also causes more solvent molecules to come in contact with solute molecules and attract them with more force. Therefore, ***more solute*** can be dissolved in the solution as the temperature gets higher due to the elevated saturation point. Such a solution is said to be ***supersaturated,*** and is unstable because the solution contains more sugar than the water can dissolve at room temperature. Solute precipitation and crystallization occur in supersaturated solutions when they are cooled down to room temperature. This process is called *recrystallization*.

Crystal formation from the solution has to do with solubility, or the largest amount of solute that can be dissolved in solvent. Rock candy is one of the best examples to explain supersaturation and crystal formation in solutions. As in Experiment 4.1, a sugar solution is supersaturated with sugar (sucrose) to make rock candy. More ***sugar than that in the saturated solution*** is dissolved in the water by heating the sugar solution. During boiling, the water evaporates and the sugar becomes even more concentrated. As the solution cools down, the sugar molecules eventually crystallize on a suitable surface for crystal formation, such as a string or pipe cleaner. Sugar crystallization can be a problem in some food products like jam and honey because they (and some desserts) are rich in sugars. Also during storage, as the moisture evaporates from the surface, the amount of sugar per volume increases and the solution gradually gets supersaturated. Eventually sugar molecules come back together as sugar crystals because supersaturated solutions are unstable in nature. Sugar crystals grow and the textures of the sweets or desserts become grainy.

The size and amount of sugar crystals primarily depends on

1. the concentration of sugar in solution;
2. the temperature at which agitation of a cooling solution is initiated;

3. Presence of foreign substances with similar crystalline structures; and
4. interfering substances.

Mixing (creaming) the hot sugar solution promotes the formation of large crystals due to rapid movement of the molecules toward each other, resulting in a grainy texture. If the creaming temperature is very low, the crystal size will be very small and the product, such as a candy, will not have the desired texture. Therefore, allowing the sugar solution to cool down to a certain temperature (38°C–54°C/100–129°F) before agitation will yield smaller crystals, and a smoother and creamier texture.

Foreign substances with similar crystalline structure in the sugar solution also may result in the formation large crystals. They act as suitable surfaces for crystal formation, which will grow to larger crystals.

Interfering substances, such as butter, milk, cream, and protein, on the other hand, physically prevent the growth of large crystals by coating the crystals and preventing one crystal growing onto another.

EXPERIMENT 4.2

OBJECTIVES

- To explain the effect of temperature on the sensory properties of sugar syrup.
- To explain sugar caramelization.

Case 1

Almond Brittle Recipe

Ingredients and Equipment

- 1 cup sugar
- ¼ cup water
- 1 cup roasted whole almonds
- Measuring cup
- 1 heavy saucepan
- Stove
- Baking paper
- Sugar thermometer
- Spoon
- Spatula

Method

1. Place the water and sugar in a heavy saucepan.
2. Place the saucepan over medium heat.
3. Place the thermometer in the pan. Be sure the thermometer does not touch the bottom of the pan.
4. Measure the temperature and record your observations in Data Table 4.4 *throughout the experiment* at given temperatures in the Data Table.
5. Cook, stirring slowly with a spoon until the sugar dissolves and the syrup turns pale golden in color. Scrape the sides of the saucepan so no sugar crystals remain on the sides of the saucepan.
6. Stop stirring when the syrup starts to boil.
7. Heat until the thermometer reaches 149°C (300°F).
8. Remove from heat immediately.
9. Stir in almonds.
10. Pour the mixture onto a baking sheet, spreading the mixture quickly into an even layer with a spatula.
11. Cool for one hour.

DATA TABLE 4.4

Temperature (°C/°F)	Appearance[a]	Color[b]	Texture[c]
50/122			
115/239			
120/248			
125/257			
135/275			
149/300			

[a] Use the terms *still, clear, bubbly* to describe the appearance.

[b] Use the terms *white, light yellow, dark yellow, brown, dark brown* to describe the color.

[c] Use the terms *smooth, soft, hard, sticky, liquid, solid* to describe the texture.

Case 2

Fudge Recipe

Materials and Equipment

- 2 cups sugar
- 2 squares (1 oz size) unsweetened chocolate
- 1 cup light cream
- 1 tablespoon butter
- Measuring cup
- 1 heavy saucepan
- Stove
- Sugar thermometer
- Wooden spoon
- Platter

Method

1. Chop the chocolate into small pieces.
2. Combine sugar, chocolate, and cream in a heavy saucepan.
3. Place the thermometer in the saucepan. Be sure the thermometer does not touch the bottom of the saucepan.

4. Cook the mixture over moderate heat. Stir constantly only until sugar and chocolate melt.
5. Cook until the mixture reaches 117°C (242°F).
6. Remove from heat and cool slightly.
7. Beat until fudge begins to harden.
8. Transfer to a buttered platter.
9. Cut into diamond-shaped pieces before fudge hardens completely.
10. Evaluate the appearance, color, and the texture; record your observations in Data Table 4.5.

Case 3

Fudge (cooked at higher temperature)

Materials and Equipment

- 2 cups sugar
- 2 squares (1 oz size) unsweetened chocolate
- 1 cup light cream
- 1 tablespoon butter
- Measuring cup
- 1 heavy saucepan
- Stove
- Sugar thermometer
- Wooden spoon
- Platter

Method

1. Complete the same stages of Case 2; at stage 5, cook until the mixture reaches 128°C (262°F) instead of 117°C (242°F).
2. Compare the texture of the products of Case 2 and Case 3.

DATA TABLE 4.5

Fudge	Appearance[a]	Color[b]	Texture[c]
Case 2			
Case 3			

[a] Use the terms *still, clear, bubbly* to describe the appearance.

[b] Use the terms *white, light yellow, dark yellow, brown, dark brown* to describe the color.

[c] Use the terms *smooth, soft, hard, sticky, liquid, solid* to describe the texture.

The Science behind the Results

Sugar (sucrose) and water are boiled to make sugar syrups, and most of the sweets and desserts (Table 4.2). The water evaporates and the sugar concentrates during boiling, so the product becomes firmer.

The sensory properties of the sweets and desserts depend on the processing temperature and the other ingredients used in the recipe.

When sugar is heated at high temperatures, it decomposes into glucose and fructose. This is followed by dehydration of sugar, in which each sugar molecule loses water and they react with each other. New flavor and color compounds are formed; sugar turns brown in color and nutty in flavor. This reaction is called *caramelization of sugar*. Caramelization gives the desirable flavor and color of many foods such as coffee, beer, brittles, a variety of candies, and confectionary. Caramelization of sucrose starts at 160°C (320°F). It becomes undesirable if the sugar is heated above the caramelization temperature because it gives a burnt smell and color to the food product.

TABLE 4.2
The Changes Occurring in Sugar Syrup during Cooking

Temperature (°C/°F)	Approximate Sugar Concentration in the Syrup (%)	Heating Stage	Properties	Example
112–116/ 233–240	85	Soft ball	Soft ball is formed and it does not hold its shape when syrup is dropped into cold water	Fudge, candy filling
118–120/ 244–248	87	Firm ball	Firm ball is formed and it holds its shape when syrup is dropped into cold water	Chewy candies (caramels)
120–130/ 248–266	92	Hard ball	Hard ball is formed when syrup is dropped into cold water	Rock candy
132–143/ 269–289	95	Soft crack	Hard, but not brittle, threads are formed when syrup is dropped into cold water	Taffy
149–154/ 300–309	99	Hard crack	Brittle and breakable threads are formed when syrup is dropped into cold water	Lollipops
160–182/ 320–359		Caramelization	Color goes from clear to brown as the temperature rises; viscosity increases	Caramels

EXPERIMENT 4.3

OBJECTIVE

To explain the natural sugar contents of foods.

Ingredients and Equipment

- 2 medium onions
- Cooking spray
- 1 skillet
- Cutting board
- Knife
- Two plates
- Spoon
- Stove

Method

1. Label one plate "raw" and the other, "caramelized."
2. Chop the onions into uniformly sized pieces.
3. Place approximately half of the onions onto the plate labeled "raw."
4. Spray the skillet with cooking spray.
5. Place the remaining onion in the skillet, and cook over medium heat while stirring.
6. Continue stirring and watch, as the onion's color turns darker and darker.
7. Remove them onto the plate labeled "caramelized" when the onions are thoroughly browned.
8. Examine the appearance, smell, flavor, and texture of the samples, and record the responses in Data Table 4.6. (See EXP 4.1–EXP 4.4.)

EXP 4.1

EXP 4.2

Exp 4.3

Exp 4.4

DATA TABLE 4.6

	Color[a]	Smell[b]	Taste[c]	Texture[d]
Raw onion				
Caramelized onion				

[a] Use the terms *white, yellow, golden,* and *golden brown* to evaluate the color of the samples.
[b] Use the terms *strong, nutty* to evaluate the smell of the samples.
[c] Use the terms *pungent, sweet* to evaluate the taste of the samples.
[d] Use the terms *crispy, soft* to evaluate the texture of the samples.

THE SCIENCE BEHIND THE RESULTS

Some vegetables, such as onions, are rich in "natural sugars" (Table 4.3). When these vegetables are being cooked, the poly- and disaccharides in their structures are broken down into smaller sugar units because the bonds between the units are disturbed by heat. As heating continues, sugar in vegetables undergoes caramelization. This is the primary reason behind why onions turn brown and develop a sweeter flavor upon dry heating.

TABLE 4.3
The Natural Sugar Content of Some Vegetables

Vegetables Serving size (g/oz)	Sugars (g/oz)
Onion: 1 medium (148 g/5.22 oz)	9/.31
Carrot: 1 carrot (78 g/2.75 oz)	5/.17
Sweet corn: 1 medium (90 g/3.17 oz)	5/.17
Sweet potato: 1 medium (90g/3.17 oz)	7/.24

EXPERIMENT 4.4

OBJECTIVES

- To explain the structural properties of starch (polysaccharides).
- To observe the effects of starch types in starch gels.

Turkish Delight Recipe

Ingredients and Equipment

- 4 cups table sugar
- 1¼ cups corn starch
- 1 cup icing sugar
- 1 teaspoon cream of tartar
- 4¼ cups water
- 1 tablespoon lemon juice
- 1 cup confectioners' sugar
- Thermometer
- 2 saucepans
- Measuring cup
- Whisk
- Pan
- Wax paper
- Knife
- Stove

Method

1. In a saucepan, combine the lemon juice, sugar, and 1½ cups water on medium heat.
2. Stir constantly until sugar dissolves.
3. Allow the mixture to boil.
4. Reduce heat to low and allow to simmer, until the mixture reaches 115.5°C (240°F).
5. Remove from heat and set aside.
6. Combine cream of tartar, 1 cup corn starch, and remaining water in another saucepan over medium heat.
7. Place the thermometer in the saucepan.
8. Stir until all lumps are gone and the mixture begins to boil.
9. Stop stirring when the mixture has a glue-like consistency. This is called the gelatinization temperature. Record the gelatinization temperature in Data Table 4.7.
10. Continue heating. Record the temperature every 10 minutes in Data Table 4.7 until the mixture becomes golden in color.

DATA TABLE 4.7

Time (min)	Temperature (°C/°F)		
	Corn starch	Potato starch	Wheat starch
0			
10			
20			
30			
40			
50			
60			
70			
...			
Gelatinization Temperature			

11. Stir in the lemon juice and water/sugar mixture.
12. Stir constantly for about five minutes.
13. Reduce heat to low; allow to simmer for one hour, stirring frequently (EXP 4.5).
14. Once the mixture has become viscous and opaque, pour the mixture into a previously wax paper or parchment-lined pan to about 1 cm (.39 in) from the top. Add nuts (optional).
15. Spread evenly.
16. Leave the sample at room temperature overnight.

EXP 4.5

DATA TABLE 4.8

Starch Type	Transparency[a]	Color[b]	Texture[c]	Change in Height (%)
Maize				
Potato				
Wheat				

[a] Use terms *transparent, translucent*, and *opaque* to describe the transparency.
[b] Use terms *white, bright golden, golden*, and *dark golden* to describe the color.
[c] Use terms *soft, semisoft*, and *firm* to describe the texture.

17. Cut in small rectangles, sprinkle with icing sugar.
18. Analyze the product and record your observations in Data Table 4.8.

Repeat above procedure using potato starch and wheat starch.

Note: In cases when the observations or the measurements cannot be done, denote that as n/a (not applicable).

To calculate the gel strength:

1. Measure the height of the Turkish Delight using a toothpick while it is still on the tray, (h_i).
2. Turn over the baking pan containing Turkish Delight onto a clean counter and remeasure the height, (h_f).

   Calculate the percentage (%) change in the height:

$$\% \text{ Change} = \frac{(h_i - h_f)}{h_i} \times 100$$

Note: A low percentage (%) change indicates a strong gel.

Study Questions

1. Compare the gelatinization temperatures of starches.
2. Discuss the gel strength of samples prepared using different starches.
3. Which starch would be more suitable to make Turkish Delight? Explain your answer.

THE SCIENCE BEHIND THE RESULTS

Starch is a polysaccharide, which is a major component of many plants and an important ingredient for the culinary industry.

Starches are the long-chain polymers of glucose units. The number of glucose molecules varies from hundreds to several hundred thousand, depending on the type of starch. Seeds, cereals, roots, and the tubers are the main sources of starch. The most common starches in culinary processes are corn (maize), rice, wheat, potato, and tapioca (cassava).

The functional properties of starches in foods are almost unlimited and the primary functions can be listed as:

- Thickening
- Body and texture formation
- Binding
- Coating
- Moisture retention

The functional properties of starch are primarily connected to its chemical composition. Starch consists of two fractions: amylose and amylopectin. These are packed into water-insoluble granules (Table 4.4). Amylose is a linear chain of glucose units. It has a simpler structure compared to amylopectin because amylopectin has a branched structure.

The relative amounts of amylose and amylopectin differ among different starch varieties depending on the plants from which they are produced (Table 4.5).

Starch is insoluble in water. It undergoes starch gelatinization when ***heated in liquid.*** When starch is mixed with water and heated, hydrogen bonds break, allowing water to enter the starch granule. Some amylose goes into the water from the surface of the starch granule; more water enters into the granule and the starch granules swell. Gradually, granules lose their natural structures. Hydrogen bonding between amylose and water are formed. The amount of free water decreases, the viscosity of the starch mixture increases. New gel structure is formed upon cooling, and the mixture thickens. This process is called *starch gelatinization* (Figure 4.1).

The chemical structure of the starch and the size of the starch granule determine the viscosity of the starch gel, the speed of gelatinization, the gel strength, and the gelatinization temperature. That is because using the correct type of starch is important in culinary processes.

TABLE 4.4
Properties of Amylose and Amylopectin

Amylose	Amylopectin
—G—G—G—G—	—G—G—G—G—G—G— (branched)
• It has a linear chain structure. • It makes approximately 10–20% of the total amount of starch in plants based on its source.	• It has a highly branched structure. • It makes approximately 80–90% of the total amount of starch in plants based on its source.

TABLE 4.5
Approximate Amounts of Amylose and Amylopectin of Various Starches

Type of Starch (source)	Amylose (%)	Amylopectin (%)
Corn (cereal)	26	74
Rice (cereal)	17	83
Potato (tuber)	21	79
Cassava (root)	17	83
Wheat (cereal)	25	75

The amylose-to-amylopectin ratio in the starch determines the characteristic properties of the final food. The amount of amylose molecules in the granule determines the gel strength. The more amylose, the stronger the gel, but the less viscous the product. This is because the amylose has a simple linear structure in which molecules can move closer to each other to form bonds. Amylopectin contributes the viscosity of the final food product. The branched structure of amylopectin molecules keeps them from coming closer to form bonds. It does not contribute to the gel formation, but it gives viscosity to the food products

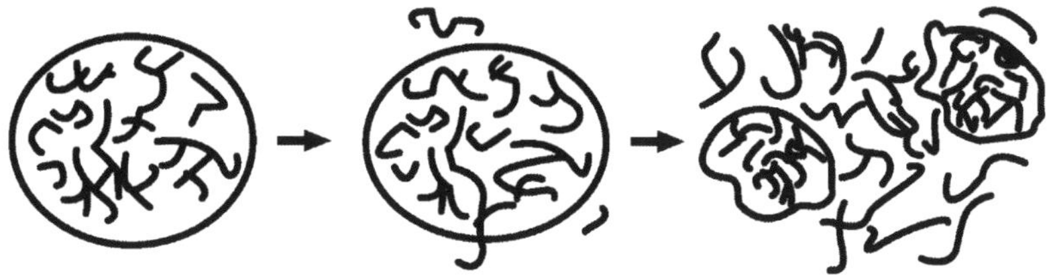

FIGURE 4.1
STARCH GELATINIZATION.

TABLE 4.6
Gelatinization Temperatures of Some Starch Varieties

Type of Starch (source)	Gelatinization Temperature (°C/°F)
Corn (cereal)	62–70/143–158
Rice (cereal)	68–75/154–167
Potato (tuber)	59–68/138–154
Cassava (root)	65–68/149–154
Wheat (cereal)	52–54/125–129

due to its bulky structure. Therefore, the more amylopectin, the softer the gel because molecules cannot align as easily and, thus, give weaker hydrogen bonding and gel strength. Different starch types have different gelatinization temperatures because they have different amounts of amylose and amylopectin in their structures (Table 4.6).

EXPERIMENT 4.5

OBJECTIVE

To explain the effects other ingredients in the recipe on starch gels.

Case 1

Sugar effect

Ingredients and Equipment

- 120 g (6.3 oz) corn starch
- 30 g (1 oz) sugar
- Water
- 6 small bowls
- 11 saucepans
- Measuring cups
- Kitchen scale
- Thermometer
- Spoon
- Whisk
- Skewers
- Refrigerator
- Stove

Method

1. Label the bowls "control," "2 g (.07 oz) sugar," "4 g (.14 oz) sugar," "6 g (.21 oz) sugar," "8 g (.28 oz) sugar," and "10 g (.35 oz) sugar."
2. Place 20 g (.7 oz) starch and 300 ml (10 oz) water in a saucepan.
3. Place the thermometer in the saucepan.
4. Heat the mixture over medium heat stirring constantly.
5. Remove from the heat when gelatinization is complete.
6. Record the gelatinization temperature in Data Table 4.9.
7. Pour the gelatinized starch paste into "control" bowl to about 2 cm (¾ in) from the top.
8. Rinse the saucepan to remove the paste residues.
9. Place 20 g (.7 oz) starch, 2 g (.07 oz) sugar, and 300 ml (10 oz) water in a saucepan.
10. Place the thermometer in the saucepan.
11. Heat the mixture over medium heat, stirring constantly.
12. Remove from the heat when gelatinization is complete.
13. Record the gelatinization temperature in Data Table 4.9.

14. Pour the gelatinized starch paste into the bowl labeled "2 g (.07 oz) sugar."
15. Store all the samples in the ***refrigerator overnight.***
16. Assess and compare the gel strengths next day.
17. Record your observations in Data Table 4.9.

Repeat steps 9 to 17 with 4 g (.14 oz) sugar, 6 g (.21 oz) sugar, 8 g (.28 oz) sugar, and 10 g (.35 oz) sugar additions. Store all the samples in the refrigerator overnight.

DATA TABLE 4.9

Sample	Gelatinization Temperature (°C/°F)	% Change in Height	Visual and Textural Properties[a]
Control			
Sugar (g/oz)			
2			
4			
6			
8			
10			
Acid (ml/oz)			
2			
4			
6			
8			
10			

[a] Use the terms ***transparent, translucent opaque, white, bright golden, golden***, and ***dark golden*** to describe the visual properties; ***soft, semisoft***, and ***firm*** to describe the textural properties.

Case 2

Acidity effect

Ingredients and Equipment

- 120 g (6.3 oz) corn starch
- 30 ml (1 oz) lemon juice
- Water
- 6 small bowls
- 1 saucepan
- Measuring cups
- Spoon
- Skewers
- Kitchen scale
- Thermometer
- Refrigerator
- Stove

Method

1. Label the bowls "control," "2 ml (.06 oz) lemon juice," "4 ml (.13 oz) lemon juice," "6 ml (.20 oz) lemon juice," "8 ml (.27 oz) lemon juice," and "10 ml (.33 oz) lemon juice."
2. Place 20 g (.7 oz) corn starch and 300 ml (10 oz) water in a saucepan.
3. Place the thermometer in the saucepan.
4. Heat the mixture over medium heat, stirring constantly.
5. Record the gelatinization temperature in Data Table 4.9.
6. Pour the gelatinized starch paste into "control" bowl to about 2 cm (¾ in) from the top.
7. Store in the *refrigerator overnight*.
8. Rinse the saucepan to remove the paste residues.
9. Place 20 g (.7 oz) corn starch, 2 ml (.06 oz) lemon juice, and 300 ml (10 oz) water in a saucepan.
10. Place the thermometer in the saucepan.
11. Heat the mixture over medium heat stirring all the time.
12. Remove from the heat when gelatinization is complete.
13. Record the gelatinization temperature in Data Table 4.9.
14. Pour the gelatinized starch paste into previously marked bowl to about 2 cm from the top.
15. Store all the samples in the *refrigerator overnight*.
16. Assess and compare the gel strengths next day.
17. Record your observations in Data Table 4.9.

Repeat steps 9 to 17 with 4 ml (.13 oz) lemon juice, 6 ml (.2 oz) lemon juice, 8 ml (.27 oz) lemon juice, and 10 ml (.33 oz) lemon juice.

Method to assess the gel strength:

1. Measure the height of the gel using a skewer while still in the bowl, (h_i).
2. Turn over bowl containing starch paste onto a clean counter and measure the height again, (h_f).

   Calculate the percentage (%) change in the height.

$$\% \text{ Change} = \frac{(h_i - h_f)}{h_i} \times 100$$

Note: A low percentage (%) change indicates a strong gel.

THE SCIENCE BEHIND THE RESULTS

Although the structure of the starch and the starch granule sizes are the primary factors, the viscosity, the speed of gelatinization, the gel strength, and the gelatinization temperature also are affected by the following factors during cooking:

1. Presence of the acidic ingredients
2. Presence of sugar
3. Presence of fats and protein
4. Presence of enzymes
5. Agitation/mixing
6. Rate of cooling

Acidic ingredients, such as lemon, yogurt, and vinegar, change the net charge in foods. They decrease the gel strength because acid breaks down the bonds between the starch molecules. Adding the acid after gelatinization can minimize the acid effect.

Sugar addition, especially sucrose addition, decreases the viscosity and firmness of the starch gel because sugar competes with the starch for water. A higher temperature of gelatinization is obtained. Timing of sugar addition is significant.

Increased concentration of *fats* and *protein* delays the hydration of starch causing a decreased viscosity of the starch paste and decreased gel strength because they coat the surface of the starch granules.

Enzymes decrease gel strength because some enzymes hydrolyze the starch. For example, an egg yolk decreases the strength of the gel because it contains alpha amylase, an amylose-digesting enzyme.

Agitation throughout the gelatinization process creates a more uniform mixture without lumps because it separates the starch granules. However, it prevents formation of a network and firm gel if the mixture is excessively agitated, or continued after the gel formation.

Rate of cooling is also important; too fast or too slow cooling may affect the gel strength.

EXPERIMENT 4.6

OBJECTIVES

- To explain the changes in the starch gel structure during storage.
- To explain the effect of storage temperature on starch gel structure during storage.

Case 1

Bread staling

Ingredients and Equipment

- A loaf of *plain* bread
- 15 medium-sized zipped sandwich bags
- Sharp bread knife
- Cutting board
- Refrigerator
- Freezer

Method

1. Label sandwich bags "room temperature," "freezer temperature," and "refrigeration temperature." Label five bags for each temperature.
2. Cut 15 *equal-sized* bread slices from the same bread.
3. Place one slice in each bag.
4. Seal the plastic bags.
5. Place five bags labeled "freezer temperature" in the freezer, five bags labeled "refrigeration temperature" in the refrigerator, and five bags labeled "room temperature" in a safe place in the room.
6. After 24 hours, take one bag from each storage temperature and let the freezer and refrigerator samples warm to room temperature.
7. Track the accumulation of water on inside surface of the plastic bags.
8. Take the slices out of the bag and evaluate the changes in the texture.
9. Compare the textures of the slices.
10. Record your evaluations in Data Table 4.10.
11. Dispose of the slices.
12. Repeat the same process over the next four days with the rest of the samples.

DATA TABLE 4.10

Storage Time	Storage Temperature	Water Accumulation on Inside Surface of Plastic Bag[a]	Texture of the Slice (Hardness, Softness)[b]	Comparison of Texture of Slices
DAY 1	Room			
	Refrigeration			
	Freezer			
DAY 2	Room			
	Refrigeration			
	Freezer			
DAY 3	Room			
	Refrigeration			
	Freezer			
DAY 4	Room			
	Refrigeration			
	Freezer			
DAY 5	Room			
	Refrigeration			
	Freezer			

[a] Use the terms *no accumulation, little accumulation, severe accumulation*.

[b] Rate the texture of each sample in a ranking of 1 to 5; 1 = hard, 5 = soft.

Case 2

Rice dishes prepared with different rice types

Ingredients and Equipment

- 100 g (3.5 oz) short grain rice (e.g., Arborio rice)
- 100 g (3.5 oz) medium grain rice (e.g., Calrose rice)
- 100 g (3.5 oz) long grain rice (e.g., Jasmine rice)
- Water
- 3 small saucepans with lids
- Spoon
- Fork
- Refrigerator
- Stove

Method

1. Label the saucepans "short grain," "medium grain," and "long grain."
2. Cook short grain rice as directed on the package.
3. When the rice is done, turn off the heat.
4. Cool down to room temperature.
5. Take off the lid.
6. Evaluate the texture.
7. Record your observations in Data Table 4.11.
8. Complete the same stages for all rice types in separate saucepans.
9. Refrigerate the saucepans overnight.
10. Take the saucepans out of the refrigerator and let come closer to room temperature.
11. Evaluate their textures.
12. Record your observations in Data Table 4.11.
13. Compare the degree of changes in textural structures of different rice dishes after refrigeration, and record in Data Table 4.11.

DATA TABLE 4.11

	Texture before Refrigeration	Texture after Refrigeration	Comparison of the Structures of Rice Dishes after Refrigeration
Short grain rice	Describe:[a] Rate:[b]	Describe: Rate:	
Medium grain rice	Describe: Rate:	Describe: Rate:	
Long grain rice	Describe: Rate:	Describe: Rate:	

[a] Use the terms *sticky*, *fluffy*, *dry*, and *moist* to describe the texture.

[b] Rate the texture of the samples in a ranking of 1 to 5; 1 = hard, 5 = soft.

The Science behind the Results

Why does bread stale during storage?
Why does cooked rice harden during storage?
Why does pudding release water during storage?

The primary cause is aging gel.

Starch molecules in gelatinized starch, particularly the amylose chains, have a tendency to reassociate in an ordered crystalline structure during storage. This process is known as retrogradation, which is directly related to the hardening of cooked rice and the staling of baked products, such as bread.

Water may squeeze out from the starch gel network as a result of retrogradation. This process is known as ***syneresis***. The collection of water on the surface of pudding during storage is a good example for syneresis.

Rate of retrogradation primarily depends on:

- Ratio of amylose and amylopectin in the starch. Due to the simpler structure of amylose, rate of retrogradation increases as the amount of amylose in the starch structure increases.
- Storage temperature has a significant effect on the rate of retrogradation. Rate of retrogradation is higher at the refrigeration temperature than the rate at the freezing temperature and at room temperature.

EXPERIMENT 4.7

OBJECTIVE

To explain the dextrinization process.

Béchamel Sauce Recipe

Ingredients and Equipment

- 3 tablespoons butter
- 2 tablespoons all-purpose flour
- 2 cups milk
- 1 teaspoon salt
- Saucepan
- Stove
- Whisk

Method

1. In a medium saucepan, melt the butter over medium-low heat.
2. Add the flour and salt and stir until smooth.
3. Cook over medium heat, stirring frequently.
4. Every two minutes, record the changes in color, taste, and the texture of the mixture in Data Table 4.12 until the mixture turns a light golden color and has a slightly nutty aroma.
5. Pour in milk slowly while whisking constantly until very smooth.
6. Observe the texture of the mixture and record in Data Table 4.13.
7. Bring to a boil.
8. Reduce the heat to low, cook the mixture 10 minutes, stirring occasionally.
9. Remove from heat.
10. Observe the texture of the mixture and record in Data Table 4.13 (EXP 4.6–EXP 4.9).

Exp 4.6

Exp 4.7

Exp 4.8

Exp 4.9

DATA TABLE 4.12

Time (min)	Color[a]	Taste[b]	Texture[c]
2			
4			
6			

[a] Use terms *white, bright golden, golden*, and *dark golden* to describe the color.
[b] Use terms *plain, nutty, light*, and *strong* to describe the taste.
[c] Use terms *silky, sandy* to describe the texture.

DATA TABLE 4.13

	Texture of the Mixture
Just after milk addition	
At the end of cooking	

Study Questions

1. Explain the effect of heat on the sensory properties of the mixture before the addition of the milk.
2. Explain the effect of heat on the sensory properties of the mixture after the addition of the milk.

The Science behind the Results

As discussed previously, heat and liquid are required for starch gelatinization. In other words, starch gels when ***heated in liquid.*** When starch is ***heated to high temperatures (100–200°C/212–392°F) without liquid*** (dry heating), the linkages between the glucose units are destroyed. Starch polymers are broken down into smaller, sweeter-tasting sugar molecules (dextrins). This process is known as *dextrinization.*

This process is applied in recipes, which call for heating of flour or starch without liquid, such as white and brown roux, some desserts (e.g., Turkish halva), and some soups. The basic purpose of the dextrinization process in the kitchen is to give a nutty aroma, sweeter taste, and a golden color to the foods. It also reduces its ability to thicken into a gel.

Note: Dextrinization may occur at lower temperatures if starch is heated with small amounts of acidic ingredients because acid helps to break down the bonds between the starch molecules.

EXPERIMENT 4.8

OBJECTIVE

To explain the basic function of pectin in foods.

Raspberry Jam Recipe

Ingredients and Equipment

- 1,000 g (2.2 lbs) of fresh raspberries
- 28 g (1 oz) powdered pectin
- 1,200 g (2.6 lbs) table sugar
- 4 tablespoons lemon juice
- Pan
- Stove
- Potato masher or a wooden spoon

Method

1. Place 500 g (1.1 lbs) of raspberries in a pan.
2. Add 2 tablespoons lemon juice.
3. Bring to a full boil over high heat, stirring constantly.
4. Using a potato masher or a wooden spoon, lightly crush raspberries.
5. Slowly add 600 g (1.3 lbs) sugar and 28 g (1 oz) of pectin while stirring constantly until the sugar is fully dissolved.
6. Return to a boil, and let it boil for five minutes.
7. Remove from the heat and let it cool down.
8. Analyze the texture and record in Data Table 4.14.
9. Repeat the experiment ***without*** pectin.
10. Analyze the texture and record in Data Table 4.14.

DATA TABLE 4.14

Sample	Texture of the Samples[a]
With pectin	
Without pectin	

[a] Use the terms ***thin, thick, viscous,*** and ***less viscous*** to evaluate the texture.

STUDY QUESTION

1. Compare the textures of the jams, and discuss the effect of pectin on the texture of the jam.

The Science behind the Results

Pectin and cellulose are the *polysaccharides* found in plants as the principal structural components of the cell walls. *They are the most abundant dietary fibers*. Fruits, vegetables, and whole grains are the most common sources of the dietary fibers.

Functional properties of fibers in culinary processes can be listed as:

- Thickening and gelling
- Emulsification
- Stabilization of emulsions
- Stabilization of foams

There are two types of fibers: water-soluble fibers and water-insoluble fibers.

Legumes, oat bran, peas, citrus fruits, apple pulp, and berries are rich in water-soluble fibers. ***Pectin*** is the most common example of the water-soluble fibers.

Cauliflower, cabbage, apple skin, wheat bran, and whole grain foods are rich in insoluble fibers. ***Cellulose*** is the most common example of the water-insoluble fibers.

Pectin is primarily used as a gelling agent, thickening agent, and stabilizer in foods, especially in jams, jellies, and some fruit juices.

It is found in plant cell walls and mostly concentrated in the skin the core of fruits. The amount of pectin varies from fruit to fruit (Table 4.7). The stage of ripeness also determines the amount of pectin in fruits. The ripe fruits contain more pectin compared to the over-ripe ones. Therefore, just-ripe fruits are used to thicken fruit syrups, jams, jellies, and marmalades. In the kitchen, fruits that are low in pectin must be combined with one of the higher pectin fruits to form a gel. Alternatively, commercial pectin can be used with low pectin fruits to form gels.

TABLE 4.7
Pectin Contents of Some Fruits

Examples of Fruits Rich in Pectin	Examples of Fruits Low in Pectin
• Sour apples	• Apricots
• Currants	• Blueberries
• Cranberries	• Sweet cherries
• Blackberries	• Peaches
• Gooseberries	• Pineapple
• Lemon	• Strawberries

EXPERIMENT 4.9

OBJECTIVE

To explain the effect of different factors on pectin gel formation.

Sour Cherry Jam

Ingredients and Equipment

- 2,000 g (4.4 lbs) sour cherries
- Water
- 1,500 g (3.3 lbs) sugar
- 10 ml (.33 oz) lemon juice
- Ladle
- Measuring cups
- Skewers
- Kitchen scale
- Large pot with a lid
- Spoons
- Timer
- 4 heat-stable jelly jars
- Thermometer
- Stove

Method

1. Label the jars "250 g (8.8 oz) sugar," "500 g (17.6 oz) sugar," "with acid," and "without acid."
2. Add 1 L (2.11 pts) of water, and 250 g (8.8 oz) of sugar into the pot.
3. Stir until the sugar is completely dissolved.
4. Wash 500 g (17.6 oz) sour cherries and add them to the pot.
5. Cook over high heat, stirring occasionally.
6. Add 5 ml (.16 oz) lemon juice.
7. Once it comes to a boil, start timing while stirring constantly.
8. Boil for 15 minutes.
9. Remove from heat.
10. Skim off any foam with a spoon.
11. Pour into a jar labeled "250 g (8.8 oz) sugar."
12. Store for a week.
13. Record your observations in Data Table 4.15.
14. Repeat the same experiment using 500 g (17.6 oz) of sugar.
15. Repeat the same experiments without acid.
16. Compare the samples.

DATA TABLE 4.15

		Flavor	Viscosity[a]	Texture[b]
Amount of Sugar (g/oz)	Amount of Acid (ml/oz)			
250/8.8	0			
250/8.8	10/.33			
500/17.6	0			
500/17.6	10/.33			

[a] Use the words *viscous, less viscous,* and *not viscous* to evaluate the viscosity.

[b] Use the words *sticky, less sticky,* and *not sticky* to evaluate the texture of the samples.

The Science behind the Results

As explained previously, thickening and gelling are the primary functional properties of pectin in culinary processes.

During processing, heat, presence of acidic ingredients, acidity of the fruit, and the amount of sugar affect the formation of a pectin gel.

Heat is required for gelling to occur. Upon heating, the pectin in fruit becomes water-soluble because heat disturbs the linkages between the molecules. Too high of a temperature or prolonged cooking can destroy the pectin, resulting in a decrease in gel strength.

Acid at a certain concentration contributes to the gel formation because it hydrolyzes pectin, and new cross links are formed. The gel will not set if there is too little acid. On the other hand, too much acid will cause the gel to lose liquid. Acid in fruit is usually enough to form the gel. In some jam recipes, especially if the fruit is not acidic, additional acid, such as lemon and cream of tartar, are required for a proper set.

Sugar has two important functions in formation of the pectin gel. The pectin cannot form a strong gel if too much water is present in the medium, which results in a runny food product. Sugar binds excessive water in fruits; therefore, the pectin can form a gel with desired texture in the jam, jellies, and the fruit syrups. Pectin also binds with sugar. Addition of acidic ingredients promotes this reaction. This is one of the reasons for using lemon juice in jam making. Acid hydrolyzes pectin, pectin binds with sugar; a gel with the proper strength is formed as a result of the synergistic effects of acid and sugar.

Note: If you are using commercial pectin, be sure to follow the instructions. Each jam, jelly, and the fruit syrup recipe must have a correct balance between pectin, acid, and sugar for the fruit you are using. Do not make unrecommended adjustments.

POINTS TO REMEMBER

- *Carbohydrates are the organic compounds naturally found in many foods, primarily in plants.*
- *Understanding the basic structure of carbohydrates is crucial for chefs because the functional properties of carbohydrates are primarily related to their structures.*
- *Carbohydrates are classified into three major groups according to the number of simple sugar (monomer) units:*
  - Monosaccharides
  - Disaccharides
  - Polysaccharides
- *Monosaccharides and disaccharides are soluble in polar solvents, such as water. Monosaccharides contain polar hydroxyl groups (OH), which can easily form hydrogen bonds with water.*
- *The sensory properties of the sweets and desserts depend on the processing temperature and the other ingredients used in the recipe.*
- *Some vegetables, such as onions, are rich in "natural sugars."*
- *Starch is a polysaccharide, which is a major component of many plants and an important ingredient for the culinary industry.*
- *Starch is insoluble in water.*
- *The chemical structure of the starch and the size of the starch granule determine the viscosity of the starch gel, the speed of gelatinization, the gel strength, and the gelatinization temperature.*
- *Starch molecules in gelatinized starch, particularly the amylose chains, have a tendency to reassociate in an ordered crystalline structure during storage. This process is known as retrogradation.*
- *There are two types of fibers: water-soluble fibers and water-insoluble fibers.*
- *Pectin is a water-soluble fiber that is primarily used as a gelling agent, thickening agent, and stabilizer in foods, especially in jams, jellies, and some fruit juices.*
- *Cellulose is the most common example of the water-insoluble fibers.*

More Ideas to Try

1. Repeat Experiment 4.7, but skip stages 3 and 4. Compare the color, taste, and the texture of the product with the final product from Experiment 4.7.
2. Repeat Experiment 4.9 by substituting apricot for sour cherry. Try it:
   a. With lemon juice
   b. Without lemon juice
   Compare the results.

Study Questions

1. Discuss the effect of sugar concentration on starch gel strength.
2. Discuss the effect of acid addition on the starch gel strength.
3. Discuss the effect of sugar concentration on pectin gel formation.
4. Discuss the effect of acid addition on pectin gel formation.

SELECTED REFERENCES

Astridottenhof, M., and I. Farhat, I. 2004. Plant biotechnology. *Biotechnology and Genetics Engineering Reviews*. 21:215–228.

Campbell-Platt, G. 2009. *Food science and technology.* Oxford, U.K.: Wiley-Blackwell Publishing Ltd.

Gaman, P. M., and K. B. Sherrington. 1996. *The science of food,* 4th ed. Oxford, U.K.: Elsevier Ltd.

Lauriston, R. 1996. Gelatinization temperatures for adjuncts. Online at: http://www.brewery.org/brewery/library/GelTemps_RL0796.html.

McGee, H. 2004. *On food and cooking,* 1st revised ed. New York: Scribner.

Murphy, P. 2000. *Handbook of hydrocolloids.* Boca Raton, FL: CRC Press.

Ozilgen, S. 2012. Failure mode and effect analysis (FMEA) for confectionery manufacturing in developing countries: Turkish delight production as a case study. *Ciência e Tecnologia Alimentos* 32(3):505–514.

Satin, M. 1998. *Functional properties of starches*. Rome: FAO Agricultural and Food Engineering Technologies Service. Online at: http://www.fao.org/ag/magazine/pdf/starches.pdf.

Spies, R. D., and R. C. Hoseney. 1982. Effect of sugar on starch gelatinization. *Cereal Chemistry* 59(2):128–131.

Starch. http://food.oregonstate.edu/learn/starch.html.

Starch. Online at: http://www.enotes.com/starch-reference/starch.

U.S. Food and Drug Administration (FDA). 2008. *Vegetables: Nutritional facts.* Online at: http://www.fda.gov/downloads/Food/LabelingNutrition/FoodLabelingGuidanceRegulatoryInformation/InformationforRestaurantsRetailEstablishments/UCM169237.pdf.

Vaclavic, V. A., and E. W. Christian. 2008. *Essentials in food science,* 3rd ed. Berlin: Springer.

CHAPTER 5

Proteins in Culinary Transformations

FUNCTIONAL PROPERTIES OF PROTEINS IN CULINARY PROCESSES

Proteins are the organic compounds naturally found in ***animals*** (e.g., milk and meat) and ***plants*** (e.g., wheat, soy, and beans).

The primary functional properties of proteins in food processing can be listed as:

1. Foam formation
2. Gelation
3. Dough formation
4. Flavor development
5. Viscosity control
6. Water binding
7. Color formation

Egg foams when whisked, but overwhisking collapses the structure. Gelatin-based jellies will not set if they are made with kiwi, but blanching the fruit solves that problem. Egg turns white when cooked, but overcooking makes it rubbery. Meat gets softer when marinated, but different marinades are used for different meat cuts—and, so on.

If we can answer the question of *why* egg albumen turns white when cooked, we will know *when* to stop the process before the egg gets rubbery. Similarly, if we know *how* meat gets softer when marinated, we will know *which* type of marinades to use for different meat cuts.

Understanding the basic structure of food proteins is crucial for chefs because the functional properties of proteins are primarily related to their structures.

PROTEIN STRUCTURE

Proteins are made up of carbon (C), hydrogen (H), oxygen (O), and nitrogen (N).

They are the polymers of different amino acids joined together by covalent bonds (peptide linkages). Each amino acid has a central carbon atom bonded to a hydrogen atom, a carboxyl group, an amino group, and *a unique functional group (R group)* (Figure 5.1). The chemical properties of R groups primarily determine the properties of proteins in foods. Therefore, proteins in foods, such as in eggs, meat, legumes, and cereals differ from each other primarily because their R groups are different.

Proteins have four structures:

1. Primary structure: The primary structure of protein is the *sequence of amino acids* bound together by peptide bonds (linkages) (Figure 5.2).
2. Secondary structure: The secondary structure of proteins is formed by coiling of the primary structure in β-sheets or α-helix shapes by formation of *hydrogen bonds* between amino acids (Figure 5.3a and Figure 5.3b).
3. Tertiary structure: The tertiary structure of proteins refers to the overall shape of protein formed by the folding of the secondary structure by

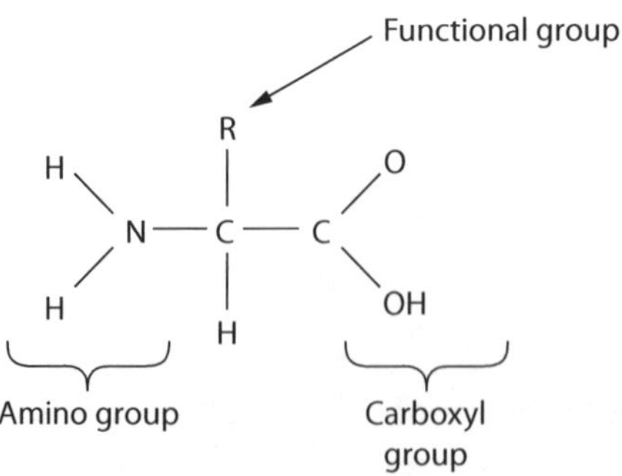

FIGURE 5.1
A GENERAL STRUCTURE OF AMINO ACIDS.

secondary attractions, i.e., disulfide bonds, hydrophobic interactions, and salt bridges (ionic interactions), between R groups of amino acids (Figure 5.4).

4. Quaternary structure: Two or more polypeptide chains join together to form the quaternary structure of proteins (Figure 5.5).

Amino acid 1

Amino acid 2

Peptide bond

$+ H_2O$

FIGURE 5.2
FORMATION OF PEPTIDE LINKAGES (BONDS).

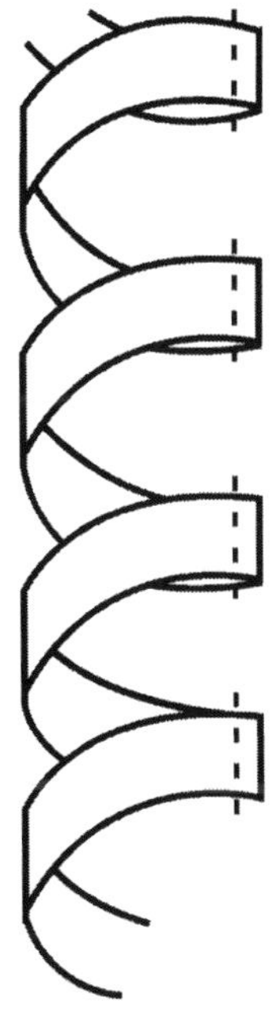

FIGURE 5.3A
α-HELIX SHAPES.

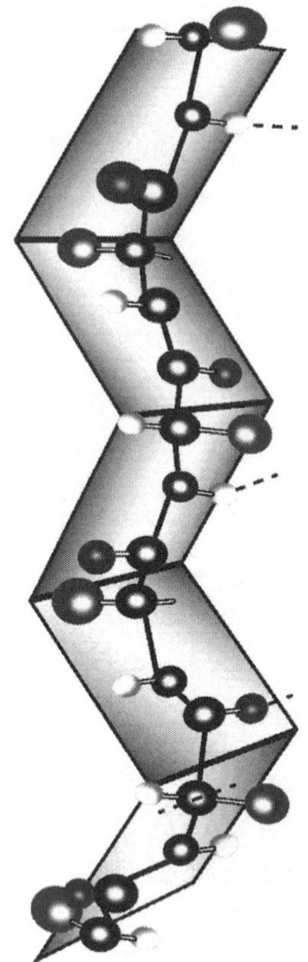

Figure 5.3B
β-Sheet.

Figure 5.4
Tertiary structure of proteins.

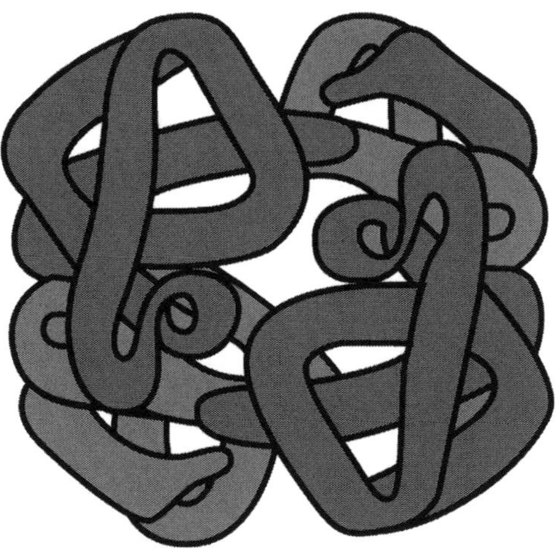

FIGURE 5.5
QUATERNARY STRUCTURE OF PROTEINS.

Different proteins in foods may have one of two different molecular shapes: globular or fibrous.

Globular Proteins	Fibrous Proteins
• Polypeptide polymers fold back on themselves as globs or spheres to form globular proteins.	• Polypeptide chains are arranged parallel to one another to form a strong rope-like structure.
• Almost all globular proteins are soluble in water.	• Fibrous proteins are insoluble in water, and are very stable.
• Milk protein (casein) and egg white protein (albumin) are good examples for the globular proteins found in foods.	• Collagen, actin, myosin, and elastin in meat are good examples of fibrous proteins found in foods.

The structure of the proteins can be affected by processing conditions.

The noncovalent bonding interactions of secondary, tertiary, and quaternary structures can be destroyed or destructed by physical or chemical treatments applied during food processing. The protein structure unfolds and a new structure is formed. This process is called ***protein denaturation*** (Figure 5.6). A new product is formed because the chemical change has taken place as a result of protein denaturation.

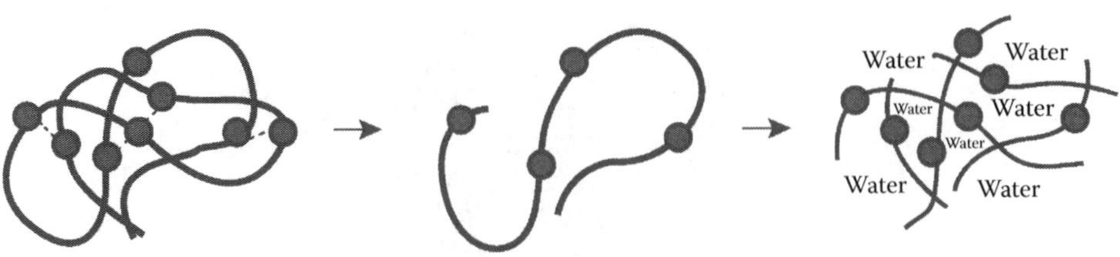

FIGURE 5.6
DENATURATION OF PROTEINS.

The most common physical and chemical factors affecting the protein structure in foods during food preparation and processing include:

- Heat treatment
- Mechanical treatment (whisking, mixing, pounding)
- Salt addition
- pH change
- Enzymatic activities
- Sugar addition

EXPERIMENT 5.1

OBJECTIVE

To explain the effect of heat treatment on proteins in foods.

Ingredients and Equipment

- 4 to 5 fresh eggs, at room temperature
- 1 tablespoon butter
- Pan
- Stove

Method

1. Heat the pan over medium heat.
2. Add butter.
3. When butter begins to sizzle, crack the eggs.
4. Observe the texture and color changes that occur in the eggs as cooking proceeds and record your observations in Data Table 5.1.
5. Cook until the egg whites have just set, and record your observations in Data Table 5.2.

DATA TABLE 5.1

Time (min)	Describe the Changes in Color[a] and Texture[b] of the Egg Whites
0	
2	
4	
6	
8	
10	
…	

[a] Use terms *transparent, translucent*, and *white* to describe the color.

[b] Use terms *liquid, semisolid*, and *solid* to describe the texture.

DATA TABLE 5.2

Color and Texture of the Egg Whites[a,b]	
Properly cooked fried eggs	
Overcooked fried eggs	

[a] Use terms *transparent, translucent*, and *white* to describe the color.

[b] Use terms *liquid, semisolid*, and *solid* to describe the texture.

6. Continue to cook for five more minutes (overcooking), and record your observations in Data Table 5.1 and Data Table 5.2.
7. Compare the textures and colors of the properly cooked eggs and the overcooked eggs (EXP 5.1–EXP 5.7).

EXP 5.1

EXP 5.2

EXP 5.3

EXP 5.4

Exp 5.5

Exp 5.6

Exp 5.7

The Science behind the Results

Understanding the composition of foods will help us understand the changes that occur within foods when we process them.

The contents of the egg is shown in PIC 5.1.

Temperature is one of the primary causes of protein denaturation during cooking. As temperature increases, the energy of the protein molecules also increases and noncovalent interactions in the protein structure are weakened. At some temperature, the protein structure unfolds. The unfolded protein molecules interact (reassociate) with each other to form a new three-dimensional network. During the reassociation of protein molecules, water is entrapped within the network and a gel-like structure is formed (Figure 5.6). The temperature at which this process occurs is known as the ***protein denaturation temperature***. Semisoft, semisolid precipitates are formed when denatured proteins are processed further. This process is called ***coagulation***. Denaturation is the first stage of coagulation. Denaturation and coagulation of proteins are usually desirable in food processing.

Differences between raw and cooked eggs are largely a result of protein coagulation that occurs during cooking.

The egg white is liquid and translucent in its raw state. Upon cooking, albumen coagulation occurs; the texture gradually changes from liquid to solid. Simultaneously, the color turns white.

If heating and denaturation continue, the interactions between proteins become very strong. The network collapses as water is removed from the network. The food becomes rubbery and dry as observed in the experiments when the egg white is overcooked.

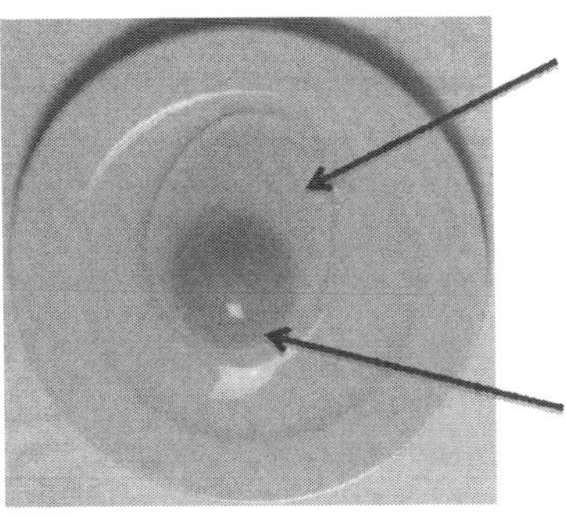

ALBUMEN (White) is the largest part of the egg. It makes up 2/3 of the egg mass. The total mass of egg white is 88% water and 12% protein.

YOLK makes up 1/3 of the egg mass. The total mass of egg yolk is 48% water, 35% lipid, and 17% protein.

Pic 5.1
Protein content of the egg.

EXPERIMENT 5.2

OBJECTIVES

- To show that different proteins have different denaturation temperatures.
- To show the effects of heating time and heating temperature on proteins in foods.

Ingredients and Equipment

- 8 eggs
- Water bath (if not available, use a large brazier with water at a constant temperature)
- 4 food thermometers
- Timer
- 16 heat-resistant glass containers

Method

1. Label four glass containers for each temperature as 60°C (140°F), 65°C (149°F), 70°C (158°F), and 75°C (167°F).
2. Take two eggs and separate the whites from the yolks.
3. Place the egg whites into two separate glass containers labeled "60°C (140°F)."
4. Repeat step 3 for the egg yolks.
5. Place the containers in a water bath (or a brazier).
6. Fill the water bath such that the water level is slightly above the level of the samples in the container.
7. Place a thermometer in each of the glass containers. Be sure that the thermometers do not touch the bottom or sides of the containers. The recorded temperatures will be the center temperature of the egg sample.
8. Heat the water in the water bath to 60°C (140°F), and keep the temperature constant.
9. Start timing when the center temperature of the sample reaches 60°C (140°F).
10. Remove one of the containers with egg white and one with egg yolk from the water bath *as soon as* the temperatures of the samples reach 60°C (140°F).
11. Evaluate the samples as described in Data Table 5.3. Record your observations and evaluations.
12. After 20 minutes, remove the other containers from the water bath.
13. Evaluate the samples as described in Data Table 5.3. Record your observations and evaluations.

Repeat the same experiment for the temperatures of 65°C (149°F), 70°C (158°F), and 75°C (167°F).

DATA TABLE 5.3

Temperature (°C)/(°F)	Appearance of the egg white[a] t = 0 min	Texture of the egg white[b] t = 0 min	Appearance of the egg yolk t = 0 min	Texture of the egg yolk t = 0 min	Appearance of the egg white t = 20 min	Texture of the egg white t = 20 min	Appearance of the egg yolk t = 20 min	Texture of the egg yolk t = 20 min
60/140								
65/149								
70/158								
75/167								

[a] Use the terms ***transparent***, ***translucent***, and ***opaque*** to describe the appearance of the samples.

[b] Use the terms ***liquid***, ***semisolid***, and ***solid*** to describe the texture of the samples both at the surface of the tube and at the center of glass custard cups.

Study Questions

1. How does cooking temperature affect the physical properties of the egg white?
2. How does cooking time affect the physical properties of the egg white?
3. How does temperature affect the physical properties of the egg yolk?
4. How does cooking time affect the physical properties of the egg yolk?

The Science behind the Results

Different proteins denature at different temperatures because they do not share the same molecular structure.

The denaturation temperature of egg whites is different from the denaturation temperature of egg yolks. Egg whites begin to denature at 62°C (143°F) and become solid at 65°C (149°F). Egg yolks begin to denature at 65°C (149°F) and become solid at 70°C (158°F). Thus, controlled cooking is required if *the mixture of the egg white and the egg yolk* are going to be used in the food preparation, such as a soufflé.

A temperature gradient occurs in the food during cooking.

The surface of the food gets hotter than the interior at the beginning of cooking. If the temperature of the cooking medium is high (intensive heating), the temperature gradient between the surface of the food and the interior will be greater; the surface cooks rapidly and gets undesirably curdled by the time the interior part gets desirably cooked. Therefore, the rate and the intensity of heating must be optimized by considering the properties of the foods and should be controlled during cooking. For example, slow cooking of the whole egg in a 60°C (140°F) water bath for 45 minutes will yield a soft, semisolid, and translucent egg white and a delicious liquid egg yolk.

EXPERIMENT 5.3

OBJECTIVE

To explain the effect of heat treatment on the textural properties of the meat.

Ingredients and Equipment

- 2 pieces of beef steaks (each 5 cm/1.9 in thick)
- 1 tablespoon olive oil
- Grill
- 2 plates
- Food brush
- Knife
- 2 food thermometers

Method

1. Preheat the grill to medium high heat (above 140°C/284°F).
2. Brush the steaks with olive oil.
3. Place the steaks on the grill.
4. Immediately place a thermometer in the center of each steak.
5. Observe the changes occurring in the texture of the samples during cooking.
6. Record the textural properties of the meats at the internal temperatures of 50°C (122°F), 60°C (140°F), and 65°C (149°F) in Data Table 5.4.
7. Take one of the steaks from the grill and place it on one of the plates while the other piece continues to cook.

DATA TABLE 5.4

Internal Temperature (°C)/(°F)	Textural Properties of the Samples[a]
50/122	
60/140	
65/149	
70/158	
75/167	

[a] Use the terms *soft, hard, juicy, dry*, and *separable* to evaluate the textures of the samples.

8. Keep observing the textural changes taking place with the second sample as it continues to cook.
9. Record the textural properties of the meat at the internal temperatures of 70°C (158°F) and 75°C (167°F) in Data Table 5.4.

Note: In this experiment, for better observation, cook only one side of the steak (EXP 5.8).

Exp 5.8

Study Questions

Slice and analyze both samples, (the sample cooked to the internal temperature of 65°C (149°F) and the sample cooked to the internal temperature of 75°C (167°F)), and answer the following:

1. Which sample is softer?
2. Which sample is juicier?
3. Which sample has more separable muscle fibers?

THE SCIENCE BEHIND THE RESULTS

Meat is composed of water, protein, fats, minerals, and trace amount of carbohydrates. Composition of lean, raw meat from different animal sources is given in Table 5.1.

Meat has muscle fibers and connective tissues, which are the naturally occurring muscle proteins found in animals.

Muscle fibers are bundles of the long and very thin cells surrounded with connective tissues. These bundles can be seen when the cooked meat is cut into pieces. They become dryer, tougher, and tear apart upon cooking.

Collagen is the major protein that makes up the connective tissues that surround the muscle fibers. It has a very strong, long structure that makes it difficult to break down. Therefore, if meat has too much collagen, it will have a tougher texture.

Textural changes are observed in meat during cooking as a result of denaturation of muscle proteins by heat.

Upon cooking, the muscle protein structure unfolds and reassociation of protein molecules occurs. The muscle fibers shrink both in diameter and in length with heat, and the water is squeezed out during association of protein molecules. As cooking progresses, collagen becomes tender, the fibrous proteins get tougher, and more moisture is lost, as is observed in Experiment 5.3.

Meat contains different proteins, which have different denaturation temperatures.

Myosin, one of the contractile filamentous proteins, shrinks both in diameter and in length at 50°C (122°F). During cooking, myosin molecules bind to each

TABLE 5.1
The Approximate Chemical Composition of Lean, Raw Meat from Different Animal Sources

Proportion (%)				
	Beef	Lamb	Chicken	Pork
Water	73.0	71.0	75.0	75.0
Protein	22.0	21.0	22.8	22.8
Fat	3.9	7.0	0.9	1.2
Other	1.1	1.0	1.3	1.0

other, thus squeezing some of the water out of the protein network. The meat becomes firmer and opaque. Raising the temperature between 62 and 65°C (143–149°F) causes shrinkage to occur more quickly, the meat suddenly releases lots of juice, and improvement of tenderness occurs due to denaturation of collagen. At temperatures of 70°C (158°F) and above, a more rapid shrinkage of collagen occurs. Prolonged heating at this temperature results in dissolution of collagen into gelatin, the muscle fibers become more easily separated, and, eventually, the meat gets more tender.

Generally, meat (notably beef) heated to an internal temperature of 60°C (140°F) is considered rare; 62 to 68°C (143–154°F), medium rare; 68 to 75°C (154–167°F), medium; and above 75°C (167°F), well done.

EXPERIMENT 5.4

OBJECTIVE

To explain how to choose the best cooking method for different cuts of meat.

Ingredients and Equipment

- 2 pieces of identical tougher meat cuts, such as meat from the neck, flank, rump, or round
- 2 pieces of identical tender meat cuts, such as meat from the loin, sirloin, or rib
- 1 tablespoon olive oil
- Water
- Grill
- Stove
- 4 plates
- 2 pots
- 2 food thermometers

Method

1. Label four plates as "tender cut cooked with dry cooking method," "tougher cut cooked with dry cooking method," "tender cut cooked with moist cooking method," and "tougher cut cooked with moist cooking method."
2. Preheat the grill to medium-high heat (above 140°C/284°F).
3. Brush one sample from each of the meat cuts with olive oil.
4. Place the meats on the grill.
5. Immediately place a thermometer in the center of each sample.
6. Cook the meat cuts on the grill until their internal temperatures read 65°C (149°F).
7. Remove the meats from the grill.
8. Place them on their respective plates, labeled "tender cut cooked with dry cooking method," and "tougher cut cooked with dry cooking method."
9. Let the samples sit for 15 minutes.
10. While the samples are sitting, fill two pots with water, and place them on the stove.
11. When the water comes to a boil, place the remaining two meat cuts into separate pots.
12. Boil the samples for 20 minutes.
13. Remove the meat cuts from the pots.
14. Place them on their respective plates, labeled "tender cut cooked with moist cooking method," and "tougher cut cooked with moist cooking method."

DATA TABLE 5.5

Sample	Texture of the Meat Cut Cooked with Dry Heating Method (Grilled)[a]	Texture of the Meat Cut Cooked with Moist Heating Method (Boiled)
Tender meat cut		
Tougher meat cut		

[a] Use the terms *soft, hard, juicy, dry* and *separable* to evaluate the texture of the samples.

15. Let them sit for 15 minutes.
16. Analyze the texture of all the samples and record your observations in Data Table 5.5.

THE SCIENCE BEHIND THE RESULTS

Muscle fibers get larger in size and muscles get stronger as an animal grows and exercises. As the muscle fibers get thicker and stronger, thicker and stronger connective tissues also are needed to bundle them. These are the major reasons why the meat from older animals, and the more active parts of animals, such as the arms and legs, are tougher compared to the meat cuts from the young and also less moving parts of the animal.

The ideal cooking method for the given meat cut is directly related to the inherent tenderness of that cut.

There are two basic methods of cooking meat: *moist heating* and *dry heating*.

Tougher cuts need to be exposed to heat for longer durations because they have more connective tissues. However, as the heat treatment gets longer, more water is removed from the meat, and the meat gets hard and dry. Therefore, tougher cuts of meat require moist and long, slow cooking, such as stewing, braising, or steaming. Tender cuts, such as loin, sirloin, or rib, are usually cooked with a dry heat method, such as roasting, grilling, or broiling.

EXPERIMENT 5.5

OBJECTIVES

- To understand the effect of heat treatment on myoglobin.
- To understand the contribution of protein to flavor, color, and textural attributes of foods.

Ingredients and Material

- 2 pieces of 5 cm-(1.9 in) thick identical beef steaks
- 1 tablespoon olive oil
- Water
- Grill
- Pot
- Stove
- 2 plates
- Food brush

Method

1. Preheat grill for medium-high heat (above 140°C/284°F).
2. Brush the first steak with olive oil.
3. Place on the grill.
4. Cook for five to seven minutes per side.
5. Transfer the steak to one of the plates.
6. Analyze and record the color, texture, smell, and taste in Data Table 5.6.
7. Put the second beefsteak in the pot.
8. Set the pot on the stove and fill it with water. The water should completely cover the beef.
9. Bring the water to a boil over medium-high heat, and then turn the heat down.
10. Simmer the beef until tender.
11. Transfer it to the second plate.
12. Analyze and record the color, texture, smell, and taste of the sample in Data Table 5.6 (EXP 5.9).

EXP 5.9

DATA TABLE 5.6

	Grilled Beef Steak	Boiled Beef Steak
Color of surface[a]		
Color of interior		
Surface texture[b]		
Texture of interior		
Smell[c]		
Taste[d]		

[a] Use the terms ***pink***, ***purple***, ***red***, and ***brown*** to evaluate the color of the samples.
[b] Use the terms *soft*, ***hard***, *juicy*, *dry*, and ***separable*** to evaluate the texture of the samples.
[c] Use the terms *fresh* and ***nutty*** to evaluate the smell of the samples.
[d] Use the terms *raw*, ***bland***, and ***nutty*** to evaluate the taste of the samples.

THE SCIENCE BEHIND THE RESULTS

Protein denaturation is one of the reasons for changes in meat color during cooking.

Myoglobin is a protein that primarily gives the color to the meat muscles. Freshly cut meat has a deep purple color. The color changes to red when the meat is exposed to oxygen. Denaturation of myoglobin occurs when the meat is heated. As temperature is increased, the color of the meat changes from red to pink at 50°C (122°F) and from pink to brown at 60°C as a result of *myoglobin denaturation*.

Another reason for changes in meat color during cooking is the Maillard browning reaction.

The Maillard browning reaction occurs between sugars and amino groups of proteins in foods, usually at higher temperatures (above 150°C/302°F.). This reaction gives the toasty and nutty flavor and smell, and the golden brown color to bread crust, fried potatoes, cookies, cakes, roasted coffee beans, and biscuits. Most consumers appreciate these changes.

The rate of the Maillard browning reaction increases as the temperature increases. Moisture slows or stops the Maillard browning reaction because the surface temperature of foods cannot exceed 100°C (212°F), especially in cooking methods where an excessive amount of water is present. That is the reason that foods roasted or grilled turn to golden brown, but foods cooked in boiling water never golden brown.

EXPERIMENT 5.6

OBJECTIVES

- To explain the effects of mechanical force on proteins in foods.
- To show the foaming properties of some proteins.
- To show the effects of sugar on foam formation.
- To show the effects of fat on foam formation.

Case 1

Meringue Recipe

Ingredients and Equipment

- 800 g (28 oz) egg whites
- 200 g (7 oz) sugar
- A pinch of salt
- Tray
- Parchment paper
- Piping bag
- Mixer
- Oven

Method

1. Preheat the oven to 110°C (230°F).
2. Line a tray with parchment paper.
3. Place 200 g (7 oz) of egg whites into a mixing bowl.
4. Add salt.
5. Beat the egg whites at high speed until they form *soft peaks.*
6. Add 100 g (3.5 oz) of sugar a spoonful at a time with the mixer running.
7. Continue to beat until the meringue forms *stiff but moist peaks.*
8. Observe the structure and record in both Data Table 5.7 and Data Table 5.8.
9. Place a spoonful of the mixture onto trays (or fill a piping bag with the mixture and squeeze out round star shapes onto the tray).
10. Place the tray into the oven and reduce the temperature to 80°C (176°F).
11. Leave the oven on for 1½ hours or until the meringues are crisp.
12. Turn the oven off and allow the meringues to cool in the oven for three to four hours.
13. Observe the structure and record in Data Table 8.

Case 2

1. Place 200 g (7 oz) of egg whites into a mixing bowl.
2. Beat the egg whites at high speed.

DATA TABLE 5.7

Case	Foam Structure[a]
1	
2	
3	
4	

[a] Use the terms *runny, soft*, and *stiff* to evaluate the foam structure of the samples.

DATA TABLE 5.8

Case 1	Texture[a]
Before cooking	
After cooking	

[a] Use the terms *runny, soft*, and *stiff* to evaluate the foam structure of the samples.

3. Continue beating for 10 more minutes after they form soft peaks.
4. Observe the structure and record in Data Table 5.7.

Case 3

1. Place 200 g (7 oz) of egg whites into a mixing bowl.
2. Add 100 g (3.5 oz) of sugar at once.
3. Beat the egg whites at high speed.
4. Observe the structure and record in Data Table 5.7.

Case 4

1. Place 200 g (7 oz) of egg whites into a mixing bowl.
2. Add one drop of egg yolk.
3. Beat the egg whites at high speed.
4. Observe the structure and record in Data Table 5.7 (EXP 5.10–EXP 5.12).

Exp 5.10

Exp 5.11

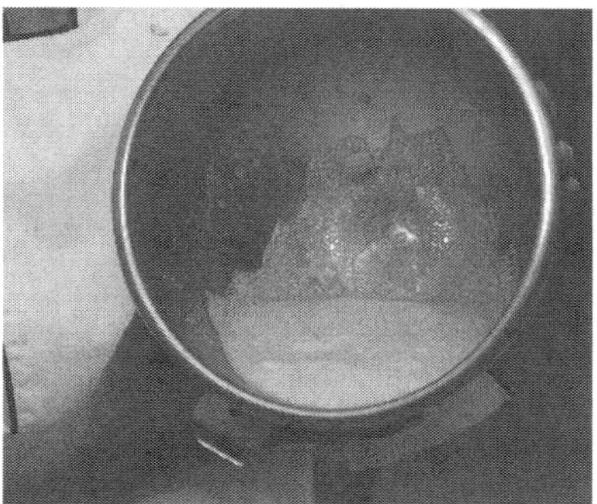

Exp 5.12

Study Questions

1. Explain what happens when the egg whites are whisked longer?
2. What is the effect of the added sugar on foam structure?
3. What is the effect of the added egg yolk on foam structure?

THE SCIENCE BEHIND THE RESULTS

When raw egg whites, cream, and milk are beaten, their protein structures unfold and the air bubbles are incorporated into the network formed by reassociation of denatured protein molecules. This process is called *foaming*. Meringue cookies and sponge cakes are good examples of foamed-structure foods.

The temperature of the egg and the environmental temperature during foam formation affect foam volume and stability.

Egg whites foam faster at room temperature. Heating or heat generation during mixing can weaken foams because heat affects the structures of proteins.

Generally, whipping time increases foaming of proteins, but intensive whipping can reduce foam stability, because the bubbles get smaller with excess mechanical force. Overmixing breaks the protein bonds in the network causing air to escape and resulting in loss of volume in the foam.

The presence of acid, salt, fat, and sugar affects the stability and ease of formation of foams.

Sugar exerts a protective effect on proteins; therefore, the rate of protein denaturation decreases as the amount of sugar increases. The addition of sugar to egg whites may decrease the volume of the foam due to its protective effect, but, on the other hand, it increases the foam stability because sugar binds the excess water. Sugar should be added *after* the formation of foam for larger foams. Addition of *salt* influences the interactions due to ionic charges. A small amount of salt promotes foam formation because it helps unfold proteins during initial foaming, but increased amounts will decrease the foam's stability by weakening the protein structure. Unlike salt, even a trace amount of *fat,* such as butter, oil, or even a drop of egg yolk will stop the foaming action of egg white proteins because fats compete with protein for special alignment with gas bubbles.

Upon cooking of the egg foam, air and/or gas inside the network heat up and expand. The protein network surrounding the air bubbles solidifies due to heat (denaturation by heat), and the firm structure is set. The denaturation temperature is elevated if an egg mixture is diluted or if it is mixed with solids, such as sugars.

EXPERIMENT 5.7

OBJECTIVE

To explain effect of acidity (pH) on proteins in foods.

Case 1

Yogurt Production

Ingredients and Equipment

- 250 ml (0.25 kg) pasteurized milk
- Yogurt culture (you can use 1 tablespoonful existing fresh yogurt)
- pH meter
- 1 pot
- 1 spoon
- 1 food thermometer
- 1 medium-size jar or a container with a lid
- Stove

Method

1. Pour the milk into a pot and heat over medium heat, stirring often until it reaches 95°C (203°F).
2. Take the milk off the stove and set it aside to cool.
3. Cool the milk to 35°C (95°F). Pour the cooled milk into a jar.
4. Measure the pH of the milk and record in Data Table 5.9.
5. Evaluate the color and consistency of the milk and record in Data Table 5.9.
6. Add one spoonful of yogurt to provide yogurt culture.
7. Mix the milk and yogurt thoroughly.
8. Cover the jar tightly and place in an incubator or an environment at 35°C (95°F) and ferment for 24 hours.
9. Slowly take the jar from the incubator, measure the pH, and evaluate the color and consistency of the content once every two hours.
10. Record your observations in Data Table 5.9.

Case 2

Protein Denaturation with Vinegar

Ingredients and Equipment

- 250 ml (8.4 oz) pasteurized milk
- 30 ml (1 oz) white vinegar
- pH meter

DATA TABLE 5.9

Time (h)	pH	Describe the Color and the Consistency of the Mixture/Yogurt[a]
0		
2		
4		
6		
8		
10		
12		
14		
16		
18		
20		
22		

[a] Use the terms *liquid, semisolid*, and *solid* to evaluate the consistency of the samples.

- 1 pot
- 1 small bowl
- 1 small plate
- 1 spoon
- A cheesecloth large enough to cover the top and 8 cm (3 in) down the sides of the pot
- 1 food thermometer
- Rubber bands
- 1 medium-size jar or a container with a lid
- Scale
- Stove

Method

1. Weigh the plate. Record in Data Table 5.10.
2. Pour the milk into a pot and heat over medium heat, stirring often until it reaches 95°C (203°F).
3. Take the milk off the stove and set it aside to cool.
4. Cool the milk to 25°C (77°F).
5. Measure the pH of the milk and record in Data Table 5.10.
6. Evaluate the color and the consistency of the milk and record in Data Table 5.10.
7. Add vinegar to the warm milk and stir for three minutes.
8. Measure the pH of the mixture.
9. Leave the milk undisturbed for 5 to 10 minutes to allow for curd formation.
10. Cover the top of the pot with the cheesecloth and fix in place using rubber bands.
11. Pour the liquid (whey) into the bowl and collect the curds in the cheesecloth.
12. Gather up the cheesecloth and squeeze out the excess whey into the bowl until almost dry.
13. Spread out the cheesecloth.
14. Place the curds on the plate and weigh the plate.

DATA TABLE 5.10

pH of the Milk	pH of the Milk + Vinegar	Weight of the Plate (g)/(oz)	Weight of the Plate + Curds (g)/(oz)	Approximate Weight of the Curds (g)/(oz)	Color and Consistency of the Milk[a]	Color and texture of the Curds[a]

[a] Use your own terms to evaluate the sample.

15. Calculate the approximate weight of the curds and record in Data Table 5.10.
16. Evaluate the color and texture of the curds and record in Data Table 5.10.

Approximate weight of the curds (g/oz) = weight of (plate + curds) (g/oz) – weight of plate (g/oz)

THE SCIENCE BEHIND THE RESULTS

The approximate composition of dairy milk can be given as 88% water, 3.3% protein, 3.3% fat, 4.7% carbohydrate, and 0.7% minerals (primarily calcium).

The primary group of milk proteins is ***casein***. Direct (as in the vinegar experiment) or indirect (as in the yogurt experiment) addition of acids and bases disturbs the structure of caseins, and casein curds are formed.

The addition of acids, bases, or certain types of salts into foods changes the net charge of the medium.

Generally, changes in the net charge of the medium disturb the bridges (bonds) in the protein structure that are held together by ionic charges, causing protein denaturation.

This reaction is the primary reason for casein curd formation in milk during yogurt and cheese manufacturing processes, and also in spoiled milk. Similarly, acid addition (such as vinegar) in the poaching liquid helps to speed up protein coagulation so that the egg white stays solid and compact.

Yogurt production is one of the most common examples used to explain the effect of acidity on protein coagulation. It is a fermented dairy product. Milk is inoculated with specific types of bacteria to make yogurt. As the bacteria grow, milk sugar (lactose) is converted into lactic acid; therefore, the pH of the medium decreases (this process is called *fermentation*). The structure of casein unfolds because the bridges in the protein structure held together by ionic charges are disrupted with the accumulation of positive charges (H^+) in the medium. The milk proteins clump together due to the charge effects and the casein curds are formed. This is known as *milk protein coagulation*.

EXPERIMENT 5.8

OBJECTIVE

To explain the effect of aging on the chemical and physical properties of eggs.

Ingredients and Equipment

- 2 half-dozen cartons of eggs
- 2 plates
- 2 mixers
- pH meter (if available)

Method

1. Purchase the first carton of eggs.
2. Mark the carton "aged."
3. Leave at room temperature for a week.
4. Purchase the second carton of eggs (preferably on the day of the experiment).
5. Mark the carton "fresh" and keep refrigerated until time for the experiment.
6. Take two plates and mark the first one "aged" and the second one "fresh."
7. Take one egg from each carton.
8. Crack the eggs onto their respective plates.
9. Observe the physical states of the egg whites and the egg yolks.
10. Record your observations in Data Table 5.11.
11. Carefully measure and record the pH of the egg white and egg yolk of each egg.

Complete the following steps for both the aged and the fresh eggs.

12. Crack and separate three eggs.
13. Place the egg whites into a mixing bowl.
14. Beat the egg whites for five minutes at high speed.
15. Record your observations in Data Table 5.12 (EXP 5.13).

Exp 5.13

DATA TABLE 5.11

	Fresh Egg	Aged Egg
Visual appearance of egg white[a]		
Visual appearance of egg yolk[a]		
pH of white		
pH of yolk		

[a] Use the terms *runny* and *viscous* to evaluate the structure of the samples.

DATA TABLE 5.12

	Fresh Egg	Aged Egg
Visual appearance before mixing[a]		
Visual appearance after mixing[a]		
Foam stability		

[a] Use the terms *runny, foamed, stiff, weak*, and *stable* to evaluate the samples.

THE SCIENCE BEHIND THE RESULTS

As the egg ages:

1. ***It loses moisture*** through the semipermeable eggshells. The air is replaced with water; therefore, the air space within the egg gets larger. The egg becomes lighter and easier to peel.
2. ***It loses carbon dioxide*** through the semipermeable eggshell. The pH of the albumen increases. The egg white becomes thinner and watery because its protein structure, which is held together by ionic charges, is disrupted with the increase in pH. It is for this reason that fresher eggs give larger and more stable foams compared to the aged eggs. The egg yolk becomes flatter and not centered because it is not protected anymore by the thick albumen. That is because pan-fried egg dishes prepared with fresher eggs have well centered egg yolks.

EXPERIMENT 5.9

OBJECTIVES

- To explain the gel formation properties of proteins.
- To explain why some fruits prevent gelatin from solidifying.
- To explain the effect of temperature on enzyme action.

Ingredients and Equipment

- 2 kiwi fruits
- 1 envelope powdered animal based gelatin (dessert gelatin)
- Water
- Ice
- 2 pots
- Large bowl
- Colander
- Skimmer
- Paper towels
- 3 dessert cups (glass)
- Cutting board
- Spoon
- Knife
- Stove
- Refrigerator

Method

1. Label the dessert cups "control," "blanched," and "fresh."
2. Peel and wash the kiwis.
3. Boil a large pot of water.
4. Slice the kiwi on a cutting board.
5. Place half of the kiwi into boiling water.
6. Blanch for two minutes.
7. Immediately place the blanched kiwi into a bowl of ice water.
8. Drain the blanched fruits using a colander.
9. Place the blanched fruits on a paper towel and pat to dry.
10. Prepare the gelatin with water according to the directions given on the package.
11. Stir well with a spoon until all gelatin is dissolved.
12. Place equal amount of gelatin mixture into each dessert cup.
13. Do not add any fruits into the "control" cup.
14. Add blanched kiwi and fresh kiwi to the corresponding cups.

DATA TABLE 5.13

Sample	Observation[a]
Control	
Gelatin dessert with fresh kiwi fruits	
Gelatin dessert with blanched kiwi fruits	

[a] Use the terms *liquid, semisolid*, and *solid* to describe the texture of the sample.

15. Refrigerate all the samples overnight.
16. After 24 hours, check the contents of each cup for solidification of the contents.
17. Record your observations in Data Table 5.13 (EXP 5.14–EXP 5.20).

Exp 5.14

Exp 5.15

Exp 5.16

Exp 5.17

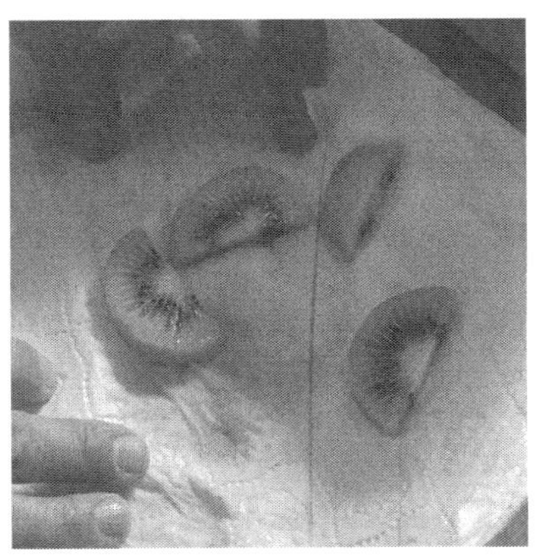

Exp 5.18

Exp 5.19

Exp 5.20

The Science behind the Results

Gelatin is a water-soluble protein prepared from collagen that is obtained from various animal sources. It is commonly used as a gelling agent in foods. Examples include desserts, ice cream, yogurt, and marshmallows.

When collagen is heated in water long enough, its structure unwinds and the chains separate, and it dissolves to form gelatin. When gelatin cools, the protein strands form a network (cross links) that can trap liquid, turning the liquid into solid gels.

The strength of a gel depends on the following:

- *Concentration of the gelatin (protein):* The higher the concentration of the protein, the stronger the gel due to increased protein–protein interactions.
- *Rate of cooling:* The higher the rate of cooling, the weaker the gel. The slower the cooling, the stronger the gel.
- *Interactions with the other ingredients in the recipe,* such as sugar, acidic ingredients, salt, or enzymes.

For example, the addition of sugar weakens the gels because sugar decreases the protein–protein association, which is required for the formation of a network. The presence of acidic ingredients and salt affect the bridges in the protein structure held together by ionic charges; therefore, a weaker gel is formed.

Proteolytic enzymes (proteases) break down the proteins into their peptides or amino acids by cleaving certain bonds in the protein structure (denaturation).

Papaya contains papain, kiwi contains bromalin, and the fig contains ficin enzymes. Therefore, using these fruits will prevent gelatin from solidifying. On the other hand, proteases also are proteins themselves. Heating of these fruits before adding them to the gelatin will allow the gelatin to solidify because heat denatures the enzymes in these fruits.

EXPERIMENT 5.10

OBJECTIVE

To explain the *synergistic effects of acids and enzymes* on proteins.

Ingredients and Equipment

- 900 ml (30 oz) pasteurized milk
- Yogurt culture (you can use 2 tablespoons of existing fresh yogurt)
- pH meter
- 1 pot
- 1 spoon
- Colander
- 1 food thermometer
- Rennet
- Cheese-pressing frame or curd knife
- Cheesecloth
- Stove

Method

1. Place the cheesecloth in the colander.
2. Pour 450 ml (15 oz) milk into the pot and heat it to 35°C (95°F).
3. Add one spoonful of yogurt to provide yogurt culture.
4. Mix the milk and yogurt thoroughly.
5. Cover the inoculated milk.
6. Let it sit in a warm place while occasionally measuring the pH of the milk.
7. Add rennet when the pH of the medium reaches close to 5 (follow the directions provided by the producers for the amount of rennet). This may take as long as eight hours.
8. Cover and leave it undisturbed for two to three hours.
9. Put your finger into the curd and lift. If the action of rennet is complete, the curd should break cleanly away and the whey should pool in the hole that has been left behind. If not, wait a little longer.
10. Cut the curds into 2- to 3-mm (0.7–0.11 in) squares. First cut parallel straight lines 2 to 3 mm (0.7–0.11 in) apart, rotate the pot 90 degrees and cut parallel straight lines 2 to 3 mm (0.7–0.11 in) apart perpendicular to the first lines. The knife or the frame must touch the bottom of the pot.
11. Place the pot over a very low heat.
12. Stir curds gently and cut larger curds.
13. When the curds are firm enough, slowly pour the them into the colander lined with cheesecloth.

DATA TABLE 5.14

Sample	Textural Properties[a]
Product prepared from milk pasteurized at 35°C (95°F)	
Product prepared from milk pasteurized at 78°C (172°F)	

[a] Use the terms *runny, semisolid*, and *solid* to describe the texture of the sample.

14. Cover the curds with the corners of the cheesecloth and lift and press the curds to remove excess whey.
15. Place it back into the colander and put a heavy weight on top.
16. Let it sit until all excess whey has been removed.

Repeat the same experiment, but heat the milk to 78°C (172°F) for 15 minutes at Step 2.

Study Questions

1. Explain your observations for the changes in the milk **with the actions of the starter culture.**
2. Explain your observations for the changes in the milk **after the addition of rennet.**
3. Compare the textural properties of the final products prepared from the milk pasteurized at 35°C (95°F) and the milk pasteurized at 78°C (172°F) (use Data Table 5.14).

THE SCIENCE BEHIND THE RESULTS

The basic principle behind natural cheese production is coagulation of milk protein with synergistic effects of acids and enzymes.

Rennet is a proteolytic enzyme obtained from the mammalian stomach. It coagulates casein in milk by cleaving particular bonds in the protein structure, causing milk to solidify.

Rennet requires specific temperatures and pH to coagulate milk.

Mild heat speeds up the enzyme reaction. The optimum temperature for milk coagulation with rennet is 35 to 42°C (95–107°F).

The optimum pH for rennet activity is around 5.8. This is one of the reasons for inoculating specific types of bacteria into milk during natural cheese production. As the bacteria grow, milk sugar (lactose) is converted into lactic acid. Therefore, the pH of the medium decreases (as in Experiment 5.7) and coagulation starts. When the pH of the milk goes down to 5.8, which is optimum for rennet activity, rennet is added in milk for further coagulation.

Note: Milk is not supposed to be heated above 65°C (149°F) to make cheese and yogurt because high temperature denatures the milk protein and coagulation will not occur.

EXPERIMENT 5.11

OBJECTIVE

To demonstrate the properties of wheat protein.

Ingredients and Equipment

- 2 cups hard wheat flour (13% protein)
- Water
- Large bowl
- Sink

Method

1. Mix water with the wheat flour.
2. Knead to make the dough structure homogenous.
3. Wash the dough under slow running cold water while kneading it throughout the process with your fingers.
4. Continue washing until the water is clear and no longer a "milky" white.
5. Squeeze the resultant ball with your fingers to remove excessive water.
6. Stretch to observe the elasticity (EXP 5.21–5.24).

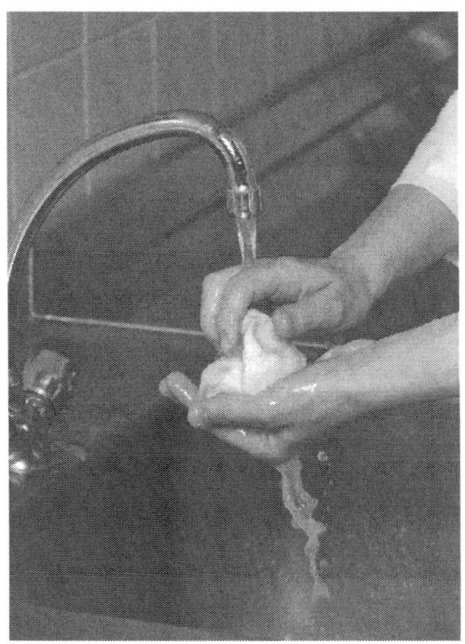

Exp 5.21

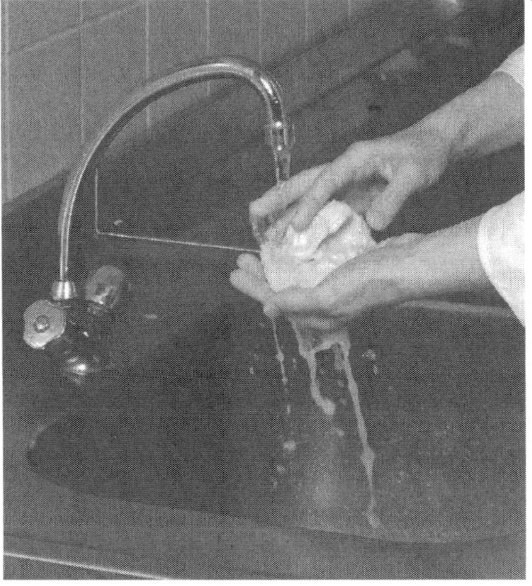

Exp 5.22

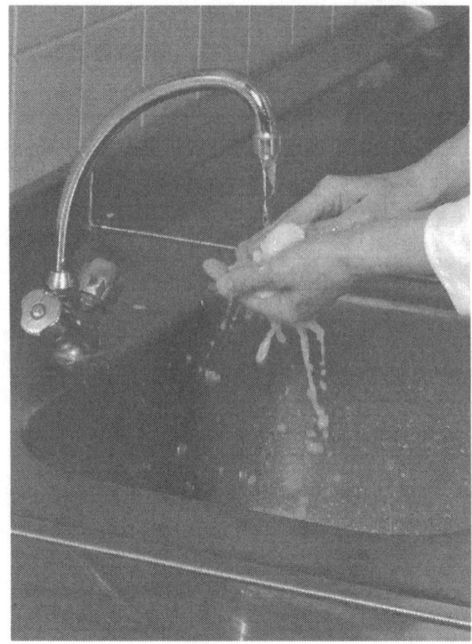

Exp 5.23

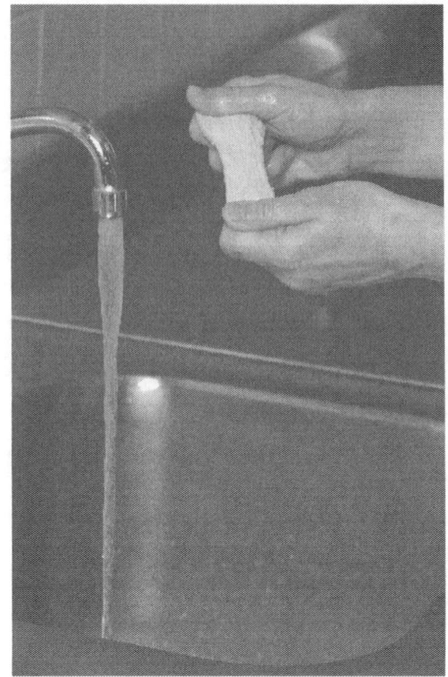

Exp 5.24

THE SCIENCE BEHIND THE RESULTS

To understand the results, we need to begin by understanding the structure of the protein in flour.

Gluten is the protein that gives the elastic structure of dough.

Gliadin and *glutenin* are the two major protein components in wheat flour. Glutenins are large proteins that provide the strength of the dough. Gliadins are smaller compared to glutenins; they ensure the elastic structure of the dough because they are more mobile due to their small size.

When water and flour are mixed, gliadin and glutenin combine to form weak gluten networks. Kneading of the dough separates these disorganized weaker protein networks, and makes the protein strands align and bind into smoother, stronger, and more organized gluten networks. Dough becomes stronger and stretchable. Inefficient kneading or excessive kneading may affect the strength of the dough because kneading ensures bond formation between the protein strands.

Other ingredients in the recipe may affect gluten development. For example, sugar and fat interfere with gluten development and the end product becomes more tender. Acidic ingredients weaken the gluten network because they change the net charge on the protein strands. On the other hand, addition of limited amount of salt helps formation of the gluten network.

In baking, carbon dioxide gas is produced by yeast or leavening agents. The carbon dioxide is trapped in the gluten network. During baking, the gas entrapped in this network expands so that the volume of the dough increases with increased temperature. As baking continues, the heat denatures the gluten proteins; a rigid framework of the baked product with small holes inside is formed.

Most of the cereal grains, such as, wheat, oats, and barley, contain gluten, but the amounts and the structures of their glutens are different.

The more gluten in the flour, the stronger and more elastic the dough becomes. The stronger structure holds more gas and expands more easily.

In the experiment, the milky white liquid that is obtained during washing contains starch granules and other water-soluble constituents.

POINTS TO REMEMBER

 *The primary functional properties of proteins in food processing include:*

- Foam formation
- Gelation
- Dough formation
- Flavor
- Viscosity control
- Water binding
- Color formation

 ***Understanding the basic structure of food proteins is crucial for chefs because the functional properties of proteins in foods are primarily related to their structures.***

 ***Proteins have four structures held together by different types of bonds.***

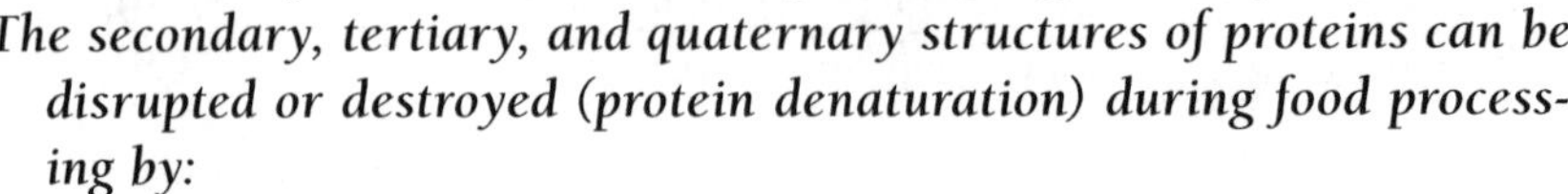 ***The secondary, tertiary, and quaternary structures of proteins can be disrupted or destroyed (protein denaturation) during food processing by:***

- Heat
- Mechanical force
- Salt
- pH
- Enzymes

 ***Different proteins may give different reactions during food processing, because they do not share the same molecular structures.***

 ***Presence or lack of the other ingredients in the recipe may affect the functional properties of the proteins.***

More Ideas to Try

1. What do you expect to observe if you repeat Experiment 5.6, Case 1, adding cream of tartar (acidic ingredient) at the fifth stage?
2. The most common application of using combined effects of acids, salts, and enzymes on protein denaturation in food preparation is meat *marination*. Marination enriches the flavor, increases the water retention, and improves the tenderness of the meat. The major ingredients used in marinades include salt solutions (2%); acidic ingredients, such as vinegar; and the enzymes, such as papain, bromelain, or enzyme-containing ingredients, such as onion. Design an experiment to show the effects of marination on the textural properties of the meat.
3. Repeat Experiment 5.11 using cookie flour and compare your observations.

Study Questions

1. Old eggs produce poor foam stability. What is the reason for this?
2. Papain and bromelain are often used as meat tenderizers. Why?
3. What is the function of acidic ingredients in meat marination?
4. Mechanical methods. such as pounding or piercing the meat, are sometimes applied to tenderize meat. What is the purpose for it?

SELECTED REFERENCES

Bradley, F. A., and A. J. King, 2004. *Egg basics for the consumer: Packaging, storage, and nutritional information.* Hollister, CA: University of California, Division of Agriculture and Natural Resources. Online at: http://anrcatalog.ucdavis.edu/pdf/8154.pdf.

Chemistry in the meat industry resources. New Zealand Institute of Chemistry. Online at: http://nzic.org.nz/ChemProcesses/animal/5A.pdf.

Crosby, G. and editors. 2012. *The science of good cooking: Master 50 simple concepts to enjoy a lifetime of success in the kitchen.* Brookline, MA: America's Test Kitchen; *Cook's Illustrated.*

Denaturation of protein. 2003. Elmhurst College. Online at: http://www.elmhurst.edu/~chm/vchembook/568 denaturation.html.

Food and Agriculture Organization. *Meat, fat and other edible carcass parts resources.* Rome: FAO. Online at: http://www.fao.org/docrep/010/ai407e/AI407E03.htm.

Forrest, J. C., E. D. Aberle,, H. B. Hedrick, and R. A. Merkel. 1975. *Principles of meat science.* New York: W. H. Freeman and Company.

Guerrero-Legarreta, I., and Y. H. Hui. 2010. *Handbook of poultry science and technology, secondary processing.* Hoboken, NJ: John Wiley & Sons.

Gustavo, G. 2009. Meat color and pH. Online at: http://www.docstoc.com/docs/122703427/Meat-color-and-pH.

Kurt, L., and S. Ozilgen. 2013. Failure mode and effect analysis for dairy product manufacturing: Practical safety improvement action plan with cases from Turkey. *Safety Science* 55:195–206.

Lomakina, K., and K. Mikova. 2006. A study of the factors affecting the foaming properties of egg white—A review. *Czech Journal of Food Sciences* 24(3):110–118.

Martens, H., E. Stabursvik, and M. Martens. 1982. Texture and colour changes in meat during cooking related to thermal denaturation of muscle proteins. *Journal of Texture Studies* 13(3):291–309.

McGee, H. 2004. *On food and cooking,* 1st revised ed. New York: Scribner.

Moore Family Center for Whole Grain Foods, Nutrition and Preventive Health. 2012. Corvallis, OR: Oregon State University, College of Public Health and Human Sciences. Online at: http://food.oregonstate.edu/learn.

Raikos, V., L. Campbell, and S. R. Euston. 2007. Effects of sucrose and sodium chloride on foaming properties of egg white proteins. *Food Research International* 40(3):347–355.

This, H. 2007. *Kitchen mysteries: Revealing the science of cooking*. New York: Columbia University Press.

Vaclavic, V. A., and E. W. Christian. 2008. *Essentials in food science,* 3rd ed. Berlin: Springer.

CHAPTER 6

Fats and Oils in Culinary Transformations

FUNCTIONAL PROPERTIES OF FATS AND OILS IN CULINARY PROCESSES

Fats and oils are the organic compounds naturally found in ***animals*** and ***plants*** (Figure 6.1).

The primary functional properties of fats and oils in food processing can be listed as:

1. *Enhancement of flavor and mouth feel*: Fats and oils provide a rich flavor and smoother mouth feel that most people find very satisfying.
2. *Development of texture*: Fats and oils make foods softer and easier to chew.
3. *Shortening*: In baking, fats and oils physically separate water and gluten molecules and provide a flaky, tender, or crumbly texture.
4. *Emulsification*: Fats and oils are the primary components of most emulsions, such as mayonnaise, certain salad dressings, sauces, and gravies.
5. *Medium for transferring heat*: During cooking, fats and oils transfer heat to the foods.
6. *Development of appearance*: Fats and oils provide a creamy, moist, fluffy, and shiny appearance to foods.

Selection of fats and oils in food processes is a critical factor because not every fat or oil is suitable for all food processing operations. For example, as butter gives

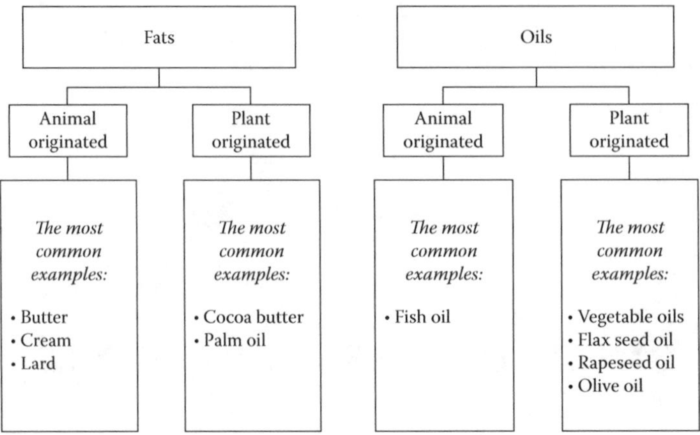

FIGURE 6.1
THE MOST COMMON FOOD SOURCES OF FATS AND OILS.

tender texture in pastry and a smooth mouth feel in breads, sunflower oil is primarily used in frying processes.

Understanding the basic structure of fats and oils is crucial for chefs because their functional properties are primarily related to their structures.

FAT AND OIL STRUCTURE

Fats and oils belong to a group of substances called lipids. They are made up of carbon (C), hydrogen (H), and oxygen (O). Lipids are most commonly the triglycerides, which are composed of one molecule of glycerol bonded to three fatty acid molecules (Figure 6.2).

All lipids are hydrophobic. This means that lipids are insoluble in water. In the figure, the R groups of fatty acids represent the long hydrocarbon chains. The hydrocarbon chains of the fatty acids can be either ***saturated*** (has no carbon–carbon double bond, C = C), ***monounsaturated*** (has one C = C), or ***polyunsaturated*** (has more than one C = C) (Figure 6.3). Naturally occurring fats and oils are the mixtures of different fatty acids in varying proportions. The degree of saturation of different lipids depends upon their various fatty acids content (Table 6.1).

The physical and chemical properties of any specific lipid primarily depend on the chemical structure of the fatty acids, such as the length of the hydrocarbon chains and degree of unsaturation that constitute it (Table 6.2).

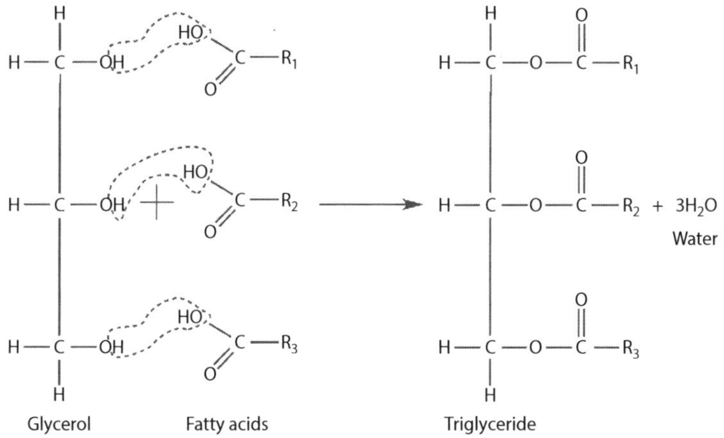

Glycerol + Fatty acids → Triglyceride + $3H_2O$ (Water)

FIGURE 6.2
FORMATION OF FATS AND OILS.

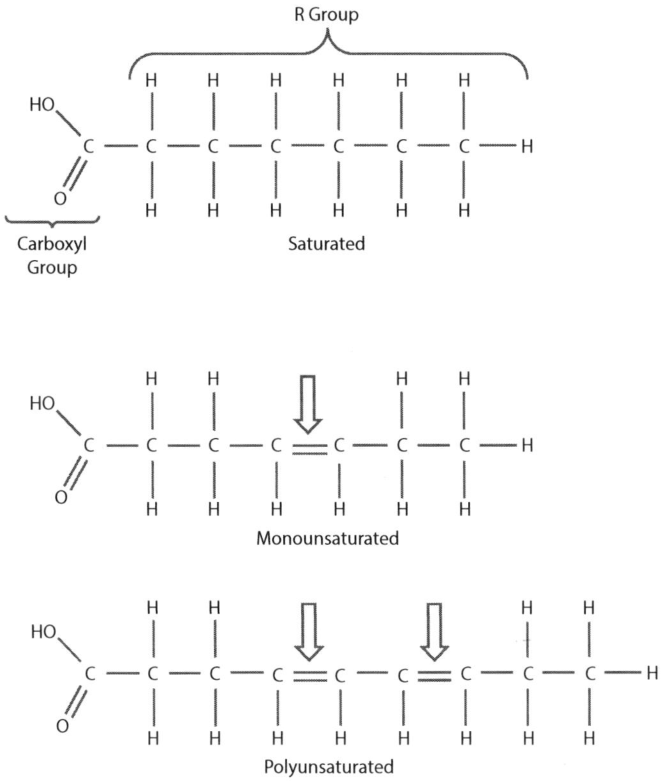

R Group

Carboxyl Group

Saturated

Monounsaturated

Polyunsaturated

FIGURE 6.3
R GROUPS OF FATTY ACIDS.

TABLE 6.1
Examples for the Saturated, Monounsaturated, and Polyunsaturated Lipids

Degree of Saturation	Example
Saturated	Butter, lard, coconut oil, tallow
Monounsaturated (MUFA)	Olive oil, rapeseed oil, canola oil, peanut oil
Polyunsaturated (PUFA)	Corn oil, cotton seed, soy bean oil

TABLE 6.2
Physical and Chemical Properties of Saturated and Unsaturated Lipids

Saturated Lipids	Unsaturated Lipids

• Hydrocarbon chains do not have double bonds between carbons.
• They are stable, and they have longer shelf life compared to unsaturated lipids.
• Their melting point is high compared to unsaturated lipids.
• They are solid at room temperature.
• They are *mostly* animal originated.
• General speaking they are known as ***fats***.
	• One or more double bonds exist in the hydrocarbon chain:
	  • ***Monounsaturated Fatty Acids*** have one double bond.
	  • ***Polyunsaturated Fatty Acids*** have two or more double bonds.
	• They are less stable, and easily undergo oxidation compared to saturated lipids.
	• They have lower melting points compared to saturated lipids.
	• They are liquid at room temperature.
	• They are mostly plant originated.
	• General speaking, they are known as ***oils***.

EXPERIMENT 6.1

OBJECTIVE

To explain smoke points of different types of lipids.

Ingredients and Equipment

- 50 ml (1.6 oz) pure olive oil
- 50 ml (1.6 oz) pure corn oil
- 50 ml (1.6 oz) pure peanut oil
- 50 ml (1.6 oz) melted clarified butter
- 50 ml (1.6 oz) melted lard
- 50 ml (1.6 oz) soybean oil
- 6 frying pans
- Liquid measuring cup
- Food thermometer
- Stove

Method

1. Pour the olive oil into a frying pan.
2. Place the thermometer in the frying pan. Position the thermometer so that the bulb is in the center of the oil and not touching the sides or bottom of the pan.
3. Heat the oil over medium-high heat until visible smoke appears from the surface of the oil.
4. Record the temperature, at which the visible smoke appears from the surface of the oil in Data Table 6.1.

Complete the same procedure for each of the fats and oils.

DATA TABLE 6.1

Type of Lipid	Smoke Point (°C)
Olive oil	
Corn oil	
Soybean oil	
Peanut oil	
Clarified butter	
Lard	

THE SCIENCE BEHIND THE RESULTS

In cooking, selecting a correct type of lipid is very important because each lipid performs best within a certain range of temperature.

The melting point of lipids can be defined as the temperature at which a solid fat becomes a liquid.

Different types of lipids have different melting point temperatures due to the differences in their chemical structures (Table 6.3). As a general rule:

- The lipids with ***longer hydrocarbon chains*** have higher melting points. In other words, the melting point increases as the ***molecular weight*** of lipids increases.
- The unsaturated lipids tend to have lower melting points than saturated lipids of the same length. In other words, the melting point of lipids decreases as the ***degree of unsaturation*** in fatty acids increases. For example, butter has a higher melting point compared to olive oil (Table 6.3). The structure of saturated fatty acids is relatively linear compared to the structure of unsaturated fatty acids. This structure allows fatty acid molecules to get closely stacked together. Therefore, close intermolecular interactions result in relatively high melting points. On the other hand, double bonds in unsaturated fatty acids result in bends in their molecular structures. Therefore, these molecules do not stack very well. The intermolecular interactions are much

TABLE 6.3
Melting Point of Selected Fats and Oils

Types of Lipid	Melting Temperature (°C/°F)
Corn oil	−11/12.2
Coconut oil	24/75.2
Olive oil	−6/21.2
Sunflower oil	−17/1.4
Soybean oil	−20/−4
Peanut oil	−2/28.4
Butter	36/96.8
Lard	33/91.4

weaker than saturated molecules. This is because the melting points are much lower in unsaturated fatty acids compared to the saturated fatty acids of the same length.

The smoke point of the lipid is one of the main properties to consider when selecting a lipid for ***high-temperature*** cooking.

Smoke point is the temperature at which lipids give off a visible bluish smoke.

Different types of lipids have different smoke points and smoke point temperatures due to the differences in their chemical structures. It is important to know the smoke point temperature of various lipids because the structure of lipids begins to decompose at that temperature (Table 6.4). The chemical decomposition results in off-flavor development, nutritional loss, and generation of harmful cancer-causing chemical components. As a general rule:

- *The saturated fatty acids* have lower smoke points than unsaturated fatty acids. For example, butter gives off a visible bluish smoke at lower temperatures compared to sunflower oil. Therefore, sunflower oil is a better choice for deep-frying processes than butter.

TABLE 6.4
Smoke Points of Selected Fats and Oils

Types of Lipid	Smoke Temperature (°C/°F)
Refined corn oil	232/449.6
Extra virgin olive oil	191/375.8
Refined sunflower oil	227/440.6
Refined soybean oil	238/460.4
Peanut oil	225/442.4
Refined canola oil	204/399.2
Coconut oil	175/348
Lard	188/370.4
Butter	121–149/249.8–300.2

- The smoke point decreases as the *free fatty acid content* increases. Prolonged or repeated heating decomposes the structure and produces free fatty acids. That is one of the reasons not to use the same oil for repeated frying processes.
- The smoke point also depends on the ***purity of the lipids***. For example, refined oils have higher smoke points compared to the unrefined ones. Presence of food particles also reduces the smoke point.

EXPERIMENT 6.2

OBJECTIVES

- To explain the importance of selecting a correct type of lipid for cooking.
- To explain that not every fat is suitable for high-temperature processes.

Ingredients and Equipment

- 1 kg (2.2 lbs) peeled potatoes
- 1 L (4¼ cups) corn oil
- 1 L (4¼ cups) peanut oil
- 1 L (4¼ cups) canola
- 1 L (4¼ cups) soybean oil
- 1 L (4¼ cups) sunflower oil
- Water
- 5 plates
- 5 medium-sized bowls
- Knife
- Skimmer
- Colander
- Measuring cup
- Paper towels
- 5 medium-sized frying pans
- Food thermometer
- Kitchen scale
- Stove

Method

1. Label the plates "corn oil," "peanut oil," "canola oil, "soybean oil," and "sunflower oil," and line them with paper towels.
2. Slice the potatoes. Make sure slices are close to the same thickness.
3. Weigh and portion the potatoes equally into 5 bowls, so that each bowl will have 200 g (7 oz) of potatoes.
4. Cover the potatoes with water.
5. Set the bowls aside for 30 minutes.
6. Place 1 L (4¼ cups) corn oil into a frying pan; 5 to 6 cm (2 to 2.3 in) of space between the top of the oil and the top of the pan is required because the oil will bubble up when potatoes are added.
7. Drain the potatoes in the first bowl and thoroughly pat dry.
8. Place the pan on the stove.
9. Heat the oil to 200°C (392°F).

DATA TABLE 6.2

Sample	Color[a]	Crunchiness[b]	Taste[b]	Smell[b]
Corn oil				
Peanut oil				
Canola				
Soybean oil				
Sunflower oil				

[a] Use the terms ***light yellow, golden***, and ***dark yellow*** to describe the color.

[b] Evaluate for crunchiness, taste, and smell, and rate the sample on a scale of 1 to 5:

1 = very soft, 5 = very crunchy

1 = less tasty, 5 = very tasty

1 = strong odor, 5 = acceptable nutty smell

10. Put the potatoes into the hot oil.
11. Fry the potatoes until they are golden brown.
12. Transfer the potatoes onto the plate labeled "corn oil."
13. Wait for one minute and remove the paper towel from the plate.
14. Evaluate the color, appearance, taste, and crunchiness of the fried potatoes and record in Data Table 6.2.

Complete the same procedure for the other oil types. Do not forget to use a separate frying pan for each type of oil.

EXPERIMENT 6.3

OBJECTIVE

To explain the factors affecting the sensory properties of fried foods.

Ingredients and Equipment

- 850 g (1.87 lbs) peeled potatoes
- 4 L (17 cups) corn oil
- Water
- Ice
- 4 plates
- 4 medium-sized bowls
- Knife
- Colander
- Measuring cup
- Paper towels
- Medium-sized frying pan
- Food thermometer
- Kitchen scale
- Stove

Method

1. Slice the potatoes. Make sure slices are close to the same thickness.
2. Weigh and portion 600 g (1.3 lbs) of the potatoes equally into three bowls so each bowl will have 200 g (7 oz) of potatoes.
3. Cover the potatoes in the first two bowls with water and cover the potatoes in the third bowl with ice.
4. Place 250 g (8.8 oz) of the potatoes into the last bowl and cover them with water.
5. Set the bowls aside for 30 minutes.

Case 1

1. Label one of the plates "control" and line it with a paper towel.
2. Drain the potatoes in the first bowl, and thoroughly pat dry.
3. Place 1 L (4¼ cups) corn oil into a frying pan; 5 to 6 cm (2 to 2.3 in) of space between

Control

Exp 6.1

the top of the oil and the top of the pan is required, because the oil will bubble up when potatoes are added.

4. Place the pan on the stove.
5. Heat the oil to 185°C (365°F).
6. Put the potatoes into the hot oil.
7. Measure the temperature of the oil.
8. Measure the time it takes to heat the oil back to 185°C (365°F).
9. Record your measurements in Data Table 6.3.
10. Fry the potatoes until they are golden brown.
11. Transfer the potatoes onto the plate.
12. Wait for one minute and remove the paper towel from the plate.
13. Evaluate the color, oil absorption, and crunchiness of the fried potatoes and record in Data Table 6.4.

DATA TABLE 6.3

Sample	Initial Temperature of Frying Oil (°C/°F)	Temperature of Oil Just after Placing Potatoes in Oil (°C/°F)	Time It Takes to Heat Oil Back to 185°C (365°F) (min)
Control	185/365		
Cold oil			
Cold potatoes	185/365		
More initial potato load	185/365		

DATA TABLE 6.4

Sample	Color[a]	Crunchiness	Oil absorption
Control			
Cold oil			
Cold potatoes			
More initial potato load			

[a] Use terms ***light yellow, golden***, and ***dark yellow*** to describe the color of the samples.
Evaluate for crunchiness and oil absorption and rate the sample on a scale of 1 to 5:
1 = very soft, 5 = very crunchy
1 = very oily, 5 = less oil absorption

Case 2

1. Label one of the plates "cold oil" and line it with paper towels.
2. Drain the potatoes in the second bowl, and thoroughly pat dry.
3. Place 1 L (4¼ cups) corn oil into a frying pan; 5 to 6 cm (2 to 2.3 in) of space between the top of the oil and the top of the pan is required because the oil will bubble up when potatoes are added.
4. Put the potatoes into the cold oil.
5. Place the pan on the stove.
6. Measure the temperature of the oil.
7. Measure the time it takes to heat the oil to 185°C (365°F).
8. Record your measurements in Data Table 6.3.
9. Fry the potatoes until they are golden brown.
10. Transfer the potatoes onto the plate.
11. Wait for one minute and remove the paper towel from the plate.
12. Evaluate the color, oil absorption, and crunchiness of the fried potatoes and record in Data Table 6.4.

Cold Oil

EXP 6.2

Cold Potatoes

EXP 6.3

Case 3

1. Label one of the plates "cold potatoes" and line it with paper towels.
2. Drain the potatoes in the third bowl, which has been covered with ice water, and thoroughly pat dry.
3. Place 1 L (4¼ cups) corn oil into a frying pan; 5 to 6 cm (2 to 2.3 in) of space between the top of the oil and the top of the pan is required because the oil will bubble up when potatoes are added.
4. Place the pan on the stove.
5. Heat the oil to 185°C (365°F).
6. Put the potatoes into the hot oil.
7. Measure the temperature of the oil.
8. Measure the time it takes to heat the oil back to 185°C (365°F).
9. Record your measurements in Data Table 6.3.
10. Fry the potatoes until they are golden brown.
11. Transfer the potatoes onto the plate.
12. Wait for one minute and remove the paper towel from the plate.
13. Evaluate the color, oil absorption, and crunchiness of the fried potatoes and record in Data Table 6.4.

Case 4

1. Label one of the plates "more potatoes" and line it with paper towels.
2. Drain the potatoes in the last bowl, and thoroughly pat dry.
3. Place 1 L (4¼ cups) corn oil into a frying pan; 5 to 6 cm (2 to 2.3 in) of space between the top of the oil and the top of the pan is required because the oil will bubble up when potatoes are added.
4. Place the pan on the stove.
5. Heat the oil to 185°C (365°F).
6. Put potatoes into the hot oil.
7. Measure the temperature of the oil.
8. Measure the time it takes to heat the oil back to 185°C (365°F).
9. Record your measurements in Data Table 6.3.
10. Fry the potatoes until they are golden brown.
11. Transfer the potatoes onto the plate.
12. Wait for one minute and remove the paper towel from the plate.
13. Evaluate the color, oil absorption, and crunchiness of the fried potatoes and record in Data Table 6.4 (EXP 6.1–EXP 6.4).

Exp 6.4

THE SCIENCE BEHIND THE RESULTS

Frying is cooking of foods in hot oils or fats. It is a popular method used in the food industry because it is possible to heat fats and oils to very high temperatures, well above the boiling point of water.

The optimal temperature for frying most foods lies between 180° to 190°C (356°F–374°F).

During frying, several physical and chemical changes occur both in foods and lipids. These changes affect the sensory properties of the end products.

The sensory properties of fried foods are primarily influenced by:

- *The type of lipids used*: During frying, a number of deteriorating chemical changes, such as structural decomposition of lipids, occur. The degree of these deteriorating chemical changes primarily depends on the chemical structure of the lipids that are used in the process. Ideal frying lipids have a high smoke point that is well above the optimal frying temperature (Table 6.4).
- *The temperature of lipids*: The frying process involves simultaneous heat and mass transfer. During food frying, when food is immersed in the hot lipid, heat is immediately transferred from the lipid to the product. As the food heats up, water vapor escapes from the food surface into the hot lipid, which results in the crust formation on the surface of the food. As frying proceeds, the crust becomes thicker and firmer. It acts as a barrier to the release of water vapor from the food, and also decreases the lipid absorption. Food becomes juicy inside and crispier outside. If the temperature of the lipid is below the optimal frying temperature or if it drops below the optimal frying temperature during the process, the crust does not form on the surface of the food. This allows lipid absorption into the food and the food becomes soft and oily.
- *The amount and the initial temperature of foods cooked*: Placing a large amount of foods in hot lipids at one time and/or starting with cold foods will decrease the temperature of the lipids below the optimal frying temperature, hence, increasing lipid absorption.
- *The processing time*: Prolonged heating will result in decomposition of lipids. Preheating the lipids any longer than required and repeated heating of the same batch of lipid must be avoided.
- *The formulation of foods*: Increased liquid content of the foods and the use of some ingredients, such as eggs and cookie flours in batters, may increase lipid absorption.

EXPERIMENT 6.4

OBJECTIVE

To explain the shortening power of different fats and oils.

Sugar Cookie Recipe

Ingredients and Equipment

- 2000 g (17.6 oz) flour
- 480 g (4.2 oz) powdered sugar
- 250 g (8.8 oz) butter
- 250 g (8.8 oz) olive oil
- 250 g (8.8 oz) margarine
- 250 g (8.8 oz) shortening
- Oven
- Tray
- Bowl
- Ruler
- Baking/parchment paper

Method

1. Preheat the oven to 160°C (320°F).
2. Mix 120g sugar with butter in a bowl.
3. Add 500g flour. Knead thoroughly to form homogeneous dough.
4. Shape the cookie dough into balls, approximately 2 cm (¾ in) in diameter each.
5. Measure the height of three balls (cookies).
6. Calculate the average height of the balls as explained in Chapter 1 and record in Data Table 6.5. h_i (Hint: The result should be close to 2 cm (¾ in).)
7. Transfer the balls onto a tray lined with baking/parchment paper.
8. Bake them in the oven for 15 minutes.
9. Cool the cookies at room temperature.
10. Measure the height of three cookies.
11. Calculate the average height of the cookies and record in Data Table 6.5 h_f.
12. Calculate the percent changes in the height of the cookies and record in Data Table 6.5.
13. Evaluate the crunchiness of the cookies and record in Data Table 6.5.
14. Evaluate the sensory properties of cookies and record in Data Table 6.5.

Complete the same procedure for the other lipid types.

DATA TABLE 6.5

Lipid	The Average Height of Balls before Cooking (cm/in)	The Average Height of Balls after Cooking (cm/in)	Crunchiness[a]	% Change in Height	Sensory Properties of Cookies after Cooking (Taste, Shape, Smell, and Color)[b]
Butter					
Olive oil					
Margarine					
Shortening					

[a] Evaluate for crunchiness of the sample on a scale of 1 to 5:
1 = very soft, 5 = very crunchy

[b] Use the ***terms round, flat,*** and ***irregularly shaped*** to evaluate the shape of the cookies. Use your own terms to evaluate the taste, smell, and the color of the cookies.

Hint:

$$\% \text{ change in the height} = \frac{(h_i - h_f)}{h_i} \times 100$$

Study Questions

1. Which type of lipid is more suitable for making these types of cookies? Explain your answer.
2. Discuss the possible reasons for the differences between the sensory properties of the cookies prepared with different lipids.

THE SCIENCE BEHIND THE RESULTS

Tenderness and flakiness are the primary quality attributes of baked food products. In batter or dough, lipids *physically* prevent a contact between water and gluten (flour protein) by coating the gluten molecules. Therefore, lipids inhibit the formation of long gluten chains in the dough. Final products become more *tender* and crumbly, the lipid is said to ***shorten*** the dough. Without shortening, the dough structure gives the feeling of hardness when chewed. This is known as *shortening power* of the lipids.

The shortening power of lipids primarily depends upon:

- *Proportion of lipid to flour*: The product gets more tender with increasing amounts of lipids.
- *Water contents of lipids*: Water in lipids may enhance the gluten development as it hydrates the medium. Therefore, the lipids with little water content have more shortening power than the lipids with high water content. Because of this, hydrogenated fats and lard have higher shortening power than butter and margarine because the later ones contain more water compared to lard and hydrogenated fats.
- *The kind of flour used*: High gluten flours produce more elastic and less tender pastry due to extensive gluten development.
- *The homogeneous distribution of the lipids* within the mixture.

Flakiness is different from tenderness. It is characterized the thin dough layers within the pastry. The lipid melts during baking and leaves empty spaces. Steam collected in these empty spaces lifts the layers of the dough.

Flakiness of the pastry primarily depends upon:

- *Degree of saturation of the lipids*: More saturated lipids produce greater flakiness than less saturated lipids because they are solid at room temperature, have higher melting points, and produce more empty spaces in the dough upon melting.
- *Temperature of the lipids*: Lipids must be well chilled to ensure that they can withstand mixing, rolling out, and handling without being creamed and or absorbed by the flour.
- *Size of the fat pieces*: Fat is cut into small pieces, but chilled to prevent it from becoming melted before cooking.
- *The kind of flour used*: High gluten flours produce flakier but tougher baked products.
- *The homogeneous distribution of the lipids* within the mixture.

EXPERIMENT 6.5

OBJECTIVES

- To explain basic structure of emulsions.
- To explain the function of an emulsifier.
- To explain the parameters affecting the emulsion structure.

Case 1: With Emulsifier

Mayonnaise Recipe

Ingredients and Equipment

- 4 egg yolks
- 1 tablespoon of vinegar
- 550 ml (2¼ cups) olive oil
- A mixing bowl
- Whisk

Method

1. Place the egg yolks in a bowl.
2. Add the vinegar and whisk to blend.
3. Add a very small amount of the oil and whisk until it's well blended.
4. Continue adding the oil while whisking thoroughly between each addition.
5. Observe the appearance and texture of the mixture when all the oil has been whisked in and record in Data Table 6.6.

DATA TABLE 6.6

Appearance and Texture of the Mixture[a]	
Mixture without emulsifier	
Mixture with emulsifier	

[a] Use the words *viscous*, *thick*, and *thin* to evaluate the appearance the texture of the mixtures.

Case 2: With No Emulsifier

Ingredients and Equipment

- 1 tablespoon of vinegar
- 550 ml (2¼ cups) olive oil
- A mixing bowl
- Whisk

Method

1. Place the vinegar in a bowl.
2. Add a very small amount of the oil and whisk until it's well blended.
3. Continue adding the oil while whisking thoroughly between each addition.
4. Observe the appearance and texture of the mixture when all the oil has been whisked in, and record in Data Table 6.6.

Case 3: Amount of Emulsifier

Ingredients and Equipment

- 2 egg yolks
- 1 tablespoon of vinegar
- 550 ml (2¼ cups) olive oil
- A mixing bowl
- Whisk

Method

1. Place the egg yolks in a bowl.
2. Add the vinegar and whisk to blend.
3. Add a very small amount of the oil and whisk until it's well blended.
4. Continue adding the oil while whisking thoroughly between each addition.
5. Observe the appearance and texture of the mixture when all the oil has been whisked in, and record in Data Table 6.7.

Case 4: Rate of Oil Addition

Ingredients and Equipment

- 4 egg yolks
- 1 tablespoon of vinegar
- 550 ml (2¼ cups) olive oil
- A mixing bowl
- Whisk

Method

1. Place the egg yolks in a bowl.
2. Add the vinegar and whisk to blend.
3. Add the oil all at once and whisk until it is well blended.
4. Observe the appearance and texture of the mixture when all the oil has been whisked in, and record in Data Table 6.7.

Case 5: A Mechanical Force

Ingredients and Equipment

- 4 egg yolks
- 1 tablespoon of vinegar
- 550 ml (2¼ cups) olive oil
- A mixing bowl
- Whisk

Exp 6.5

Method

1. Place the egg yolks in a bowl.
2. Add the vinegar and whisk to blend.
3. Add a very small amount of the oil and whisk until it's well blended.
4. Continue adding the oil while whisking thoroughly between each addition.
5. Continue whisking 10 more minutes when all the oil has been whisked in.
6. Observe the appearance and texture of the mixture, and record in Data Table 6.7 (EXP 6.5).

DATA TABLE 6.7

	Appearance and Texture of the Mixture[a]
Amount of emulsifier	
Rate of oil addition	
Mechanical force	

[a] Use the words *viscous, thick,* and *thin* to evaluate the appearance and the texture of the mixtures.

The Science behind the Results

Oil and water are immiscible liquids, meaning they do not mix. They do not mix since oil is nonpolar (hydrophobic) and water is polar. When oil and water are mixed and then left to stand, they will separate into two layers. The oil will float above the water because it is less dense than water.

Mixtures of two or more liquids that are normally immiscible are called emulsions. In emulsions, fine droplets of one liquid are dispersed in another liquid. Lipids and water are the common liquids in food emulsions. Emulsions contain three components:

1. *A dispersed phase*: A liquid that is suspended in the form of fine droplets in a continuous phase.
2. *A continuous phase*: A liquid phase that contains fine droplets of the other liquid (dispersed phase). The volume of the continuous phase is larger than that of the dispersed phase.
3. *The emulsifier*: A substance that holds water and oil in place in emulsions. Emulsifiers have both polar (hydrophilic) and nonpolar (hydrophobic) ends. In the mixture, the polar end of the emulsifier attracts the water and the nonpolar end attracts the oil. It forms a film at the boundary of oil and water, which coats the droplets of the dispersed phase. Therefore, the oil droplets stay suspended in the water or the water droplets stay suspended in the oil phase. Emulsifiers are usually produced from natural sources. Some food ingredients are used as emulsifiers. For example, egg yolk contains lecithin, which is known to be a good emulsifier. In the emulsion, if the water is the continuous phase and the oil is the dispersed phase, the emulsion is called an oil-in-water ***(o/w)*** emulsion. Milk is a good example of natural oil-in-water emulsions, with fat droplets dispersed in a larger amount of the water phase. Mayonnaise and certain salad dressings are the other common examples of water-in-oil (o/w) emulsions. If the oil is the continuous phase and the water is the dispersed phase, the emulsion is said to be a water-in-oil ***(w/o)*** emulsion. Butter, cream, and margarine are examples of w/o emulsions.

Stable emulsions do not separate under normal handling and storage conditions. The major factors affecting emulsion stability can be listed as:

- *The average droplet size of the dispersed phase*: Larger droplets have a tendency to coalesce and form a separate phase.
- *The density differences between the continuous and dispersed phases*: Emulsions are more stable when density differences between the phases are small. There is a possibility of phase separation when the difference

between the densities of the phases is large because the less dense phase has a tendency to rise to the surface through the emulsion.

- *The viscosity of the continuous phase*: The mobility of the droplets in the emulsion decreases with increased viscosity. Therefore, emulsions are more stable when the viscosity of the continuous phase is high.
- *Type and amount of the emulsifier*: To form stable emulsions, the surface of all the droplets must be coated with an emulsifier.
- *Acidity*: Changing the pH of the medium may decrease the stability of emulsions by changing the net charge on the emulsions. As explained previously, the *attraction* between the emulsifier and the polar and nonpolar phases forms the emulsions. Changes in the charge of the emulsions may break down this attraction. For example, acid addition may break down the emulsion stability because acids provide high amounts of positively charged hydrogen ions (H^+).
- *Ionic strength*: Addition of salts may decrease the stability of emulsions. Some salts drastically change the net charge of the emulsions when they are dissolved in one of the phases.
- *Temperature*: Heating or cooling of emulsions affects the emulsion stability. Heating may cause phase separation because the oil droplets melt and coalesce when heated. Upon freezing, destabilization of emulsions occurs because the ice crystals that are formed during freezing physically disturb the film between the phases.
- *Extended storage time*: Over time, droplets may combine to form a larger droplet, so the average droplet size increases and phase separation occurs. Extended storage of the emulsified foods should be avoided.
- *Mechanical force*: Violent shaking or extended mixing may physically disturb the interfacial film formed by the emulsifier and break the emulsion.

EXPERIMENT 6.6

OBJECTIVE

To explain the effects of storage temperature and light on the oxidation of lipids.

Cornmeal Cookie Recipe

Case 1: Different Storage Conditions

Ingredients and Equipment

- 2/3 cup all-purpose flour
- 1/4 cup cornmeal
- 2 tablespoons corn starch
- 1/4 teaspoon salt
- 1/2 cup softened butter
- 1/3 cup sugar
- 1/2 teaspoon pure vanilla extract
- Measuring cup
- 2 bowls
- Cookie cutter
- 3 dark-colored jars
- 1 transparent cookie jar
- Baking/parchment paper
- Rolling pin
- Mixer
- Tray
- Oven

Method

1. Label dark-colored jars "room temperature," "refrigeration temperature," and "oven at 40°C (104°F)," respectively.
2. Label the transparent jar "exposed to light at room temperature."
3. Preheat the oven to 175°C (347°F).
4. Mix the flour, cornmeal, corn starch, and salt together.
5. Blend the butter, sugar, and vanilla until they are creamy.
6. Add all of the flour mixture into the butter mixture and mix until the dough just begins to come together.
7. Roll out the dough on a well-floured surface to 1 cm (⅓ in) thick.

8. Using a cookie cutter, cut out 16 cookies, each with a 5-cm (1.9-in) diameter.
9. Transfer the cookies to a tray lined with baking/parchment paper.
10. Bake them until they become pale golden in color.
11. Let them cool down at room temperature.
12. Place four cookies in each jar.
13. Store each jar according to the conditions marked on the label.
14. Take one cookie from each jar and evaluate the aroma and taste of the cookies.
15. Record the results in Data Table 6.8.
16. Repeat the sensory taste every two days.

DATA TABLE 6.8

Storage Conditions				
Storage Time (days)	Dark-Colored Jar, Room Temp.	Dark-Colored Jar, Refrigeration Temp.	Dark-Colored Jar, 40°C/104/°F	Transparent Jar, Exposed to Light
	Aroma[a]/Taste[a]	Aroma/Taste	Aroma/Taste	Aroma/Taste
2				
4				
6				
8				

[a] Describe and rate the aroma and taste of the samples on a scale of 1 to 5:
1= smells normal, no rancid smell, 5 = strong rancid smell
1= no rancid/bitter taste, 5 = strong rancid/bitter taste

The Science behind the Results

During storage of lipids and food products rich in lipids, development of off-odor and off-flavors may occur as a result of deteriorating changes in the structures of lipids. This phenomenon is called *rancidity* or *lipid oxidation*.

Lipid oxidation is generally classified into two broad classes:

- *Enzymatic rancidity*: This type of rancidity occurs when the enzymes naturally present in the foods break down lipids into fatty acids and glycerol. For example, lipoxygenase enzymes in soybeans may cause development of off-odor and off-flavors in food products that contain soybeans.
- *Oxidative rancidity*: This type of rancidity occurs when oxygen combines with the fatty acids in the lipids. Oxidative rancidity involves a series of complex reactions (Figure 6.4). When lipids are exposed to various factors, such as heat, their structures are degraded and highly reactive compounds that are called *free radicals* are formed. Free radicals combine with environmental oxygen and give further chain reactions. Compounds responsible for the characteristic unpleasant smell and flavor associated with oxidative rancidity are formed as a result of these reactions.

The various factors that affect the development oxidative rancidity in foods can be listed as:

- *The degree of unsaturation of fatty acids*: Unsaturated lipids are more prone to oxidative rancidity than saturated lipids. For example, nuts, peanuts, and whole wheat products are more prone to rancidity compared to butter.
- *Oxygen, heat, light, metal, and moisture*: They increase the rate of oxidative rancidity.

Several precautions that can be taken to reduce the rate of oxidative rancidity in foods include:

- Storing the sensitive foods (the foods that are prone to oxidative rancidity) in dark places.
- Reducing the exposure of sensitive foods to direct light (including sunlight).
- Packaging the foods in selectively light-absorbent packaging materials, such as dark-colored bottles.
- Keeping sensitive foods away from heat.
- Storing highly sensitive foods at refrigeration temperature.

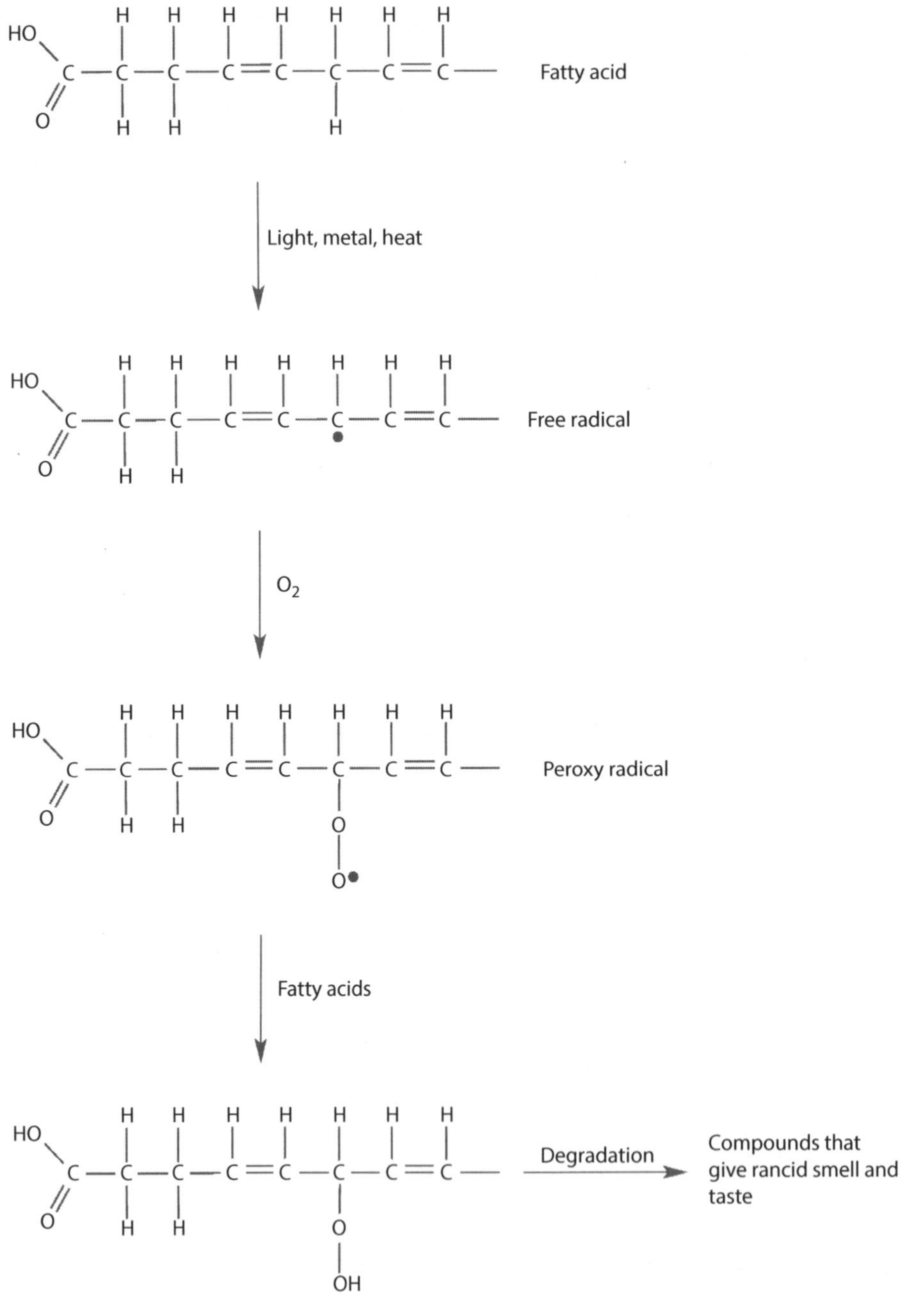

FIGURE 6.4
MECHANISM OF OXIDATIVE RANCIDITY.

- Reducing the environmental oxygen concentration to very low levels, such as vacuum packaging the sensitive food products.
- Avoiding direct contact of foods with copper and iron equipment and containers.
- Using approved natural (i.e., ascorbic acid, tocopherol) or synthetic (BHA, BHT) antioxidants.

EXPERIMENT 6.7

OBJECTIVE

To demonstrate the effects of fat crystals on sensory attributes of some food products.

Ingredients and Equipment

- 3 blocks of untempered chocolate couvertures [high-quality chocolate] (2.5 kg (5.5 lbs) each)
- 1 milk chocolate
- Water
- 2 heat-stable bowls
- 1 metal bowl
- 3 medium-sized containers (they will be placed over the top of the heat-stable bowls; therefore, the size of the containers must be chosen accordingly)
- 1 container (this will be placed in the metal bowl; therefore, the size of the container must be chosen accordingly)
- Stretch film (plastic wrap)
- Thermometer
- Spatula
- Stove
- Refrigerator

Method

Case 1: Fat Blooming

1. Label three medium containers "Tempered," "Untempered," and "Temperature."
2. Place one of the couvertures in the container labeled "Untempered."
3. Store at constant temperature of 16 to 17°C (60.8 to 62.5°F).
4. Break one of the chocolate couvertures into small pieces.
5. Put them in the container labeled "Tempered."
6. Place a small amount of water into one of the heat-stable bowls.
7. Place the container over the top of the heat-stable bowl. The bottom of the container should not touch the water.
8. Heat the water until it is very gently simmering, not boiling.
9. Check the temperature of the chocolate continuously.
10. When the temperature of the chocolate reaches 50°C (122°F) on a thermometer, remove the container from the bowl.
11. Stir the chocolate with a spatula.

12. As soon as the temperature cools to 26°C (78.8°F), return the container to the bowl and reheat.
13. Heat the chocolate, stirring gently.
14. When the temperature of the chocolate reaches 32°C (89.6°F) on a thermometer, remove the container from the bowl.
15. Store with the "untempered" sample.
16. Break the last chocolate couverture into small pieces.
17. Place the chocolate pieces into the container labeled "Temperature."
18. Place a small amount of water into the heat-stable bowl.
19. Place the container over the top of the heat-stable bowl.
20. Heat the water until it is very gently simmering, not boiling.
21. Remove the container from the bowl once the chocolate appears to have melted.
22. Cover the container with paper towel and store with the other two samples.
23. Next day, take the sample from the storage area and place it in a warm place that has a temperature of 30 to 35°C (86 to 95°F), for a few hours.
24. Return the container back to its storage area.

Case 2: Sugar Blooming

Meanwhile,

1. Place the milk chocolate into a container labeled "humidity."
2. Put *a small amount* of water into the metal bowl.
3. Place the container inside the metal bowl.
4. Cover them tightly together with stretch film (plastic wrap). That will create a humid environment.
5. Store it with the other samples.

Check the surface of the samples every day. Continue the observation for 5 to 10 days. Record your observations in Data Table 6.9 (EXP 6.6).

Note: In this experiment, a special technique, which is called a ***bain marie technique***, is applied to melt the chocolate.

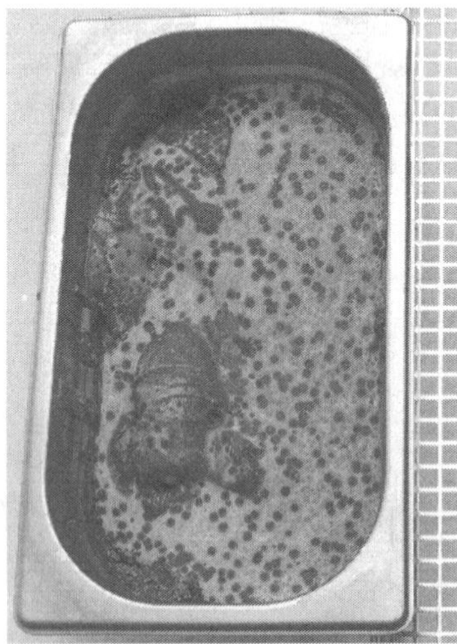

EXP 6.6

SURFACES OF UNTEMPERED CHOCOLATE STORED UNDER FLUCTUATING TEMPERATURES.

DATA TABLE 6.9

Surface Appearance[a]				
Storage Time (days)	Tempered Sample	Untempered Sample	Sample Exposed to Fluctuating Temps.	Sample Stored in a Humid Environment
0				
1				
2				
3				
4				
5				
6				
7				
...				

[a] Use the terms *no change, sign of white spots,* and *cloudy* to evaluate the surface of the samples.

The Science behind the Results

White, cloudy-looking coating, "bloom," develops on the surface of chocolates if chocolates are processed, stored, or served improperly.

There are two main types of chocolate blooms:

1. Fat bloom
2. Sugar bloom

Fat bloom is primarily related to the crystallization of cocoa butter found in chocolate products on the surface of chocolate. Cocoa butter is composed of lots of different fat crystals. Some of the crystals are stable, but other crystals are not. Unstable crystals change form over time and form a cloudy-looking coating on the surface of the chocolate. During chocolate manufacturing, a process called *tempering* is used to ensure that only the stable crystals are formed in the final chocolate products. *Tempering* is a directed precrystallization process to promote the formation of the stable crystal forms only.

Tempering is a three-step process:

Step 1: Melting and holding the chocolate at 50°C (122°F) for a sufficient time.
Step 2: Rapid cooling to about 26 to 28°C (78.8 to 82.4°F).
Step 3: Warming to 31 to 32°C (87.8 to 89.6°F).

The primary reasons for fat bloom on the surface of chocolate include the following:

- *Improper tempering of the chocolate*: If chocolate is not tempered or if the tempering process is not carried out properly, the unstable forms of cocoa butter crystals will form and fat bloom will occur on the surface of the chocolate during storage.
- *High temperature*: The chocolate should be stored below the melting point of cocoa butter to avoid surface fat bloom.
- *Temperature fluctuation*: During storage, temperature fluctuation must be avoided. The chocolate is best when stored in an area of stable temperature (15–18°C/59–64.4°F) to avoid fat blooming.

Sugar bloom is caused by moisture condensing on the chocolate. Chocolate is usually composed of ground cocoa beans, sugar, and sometimes emulsifiers and milk. When chocolate is exposed to moisture, water dissolves the sugar on the surface of the chocolate. The dissolved sugar crystallizes on the surface of the chocolate as the water dries. The resulting small sugar crystals give the chocolate a cloudy appearance. For example, if the chocolate is placed in the refrigerator and then removed and placed at room temperature, moisture from the air will condense on the cold chocolate, and, as water dries, the sugar bloom will appear on the surface of the chocolate. Sugar bloom also may occur if the chocolate is stored in an environment with very high humidity.

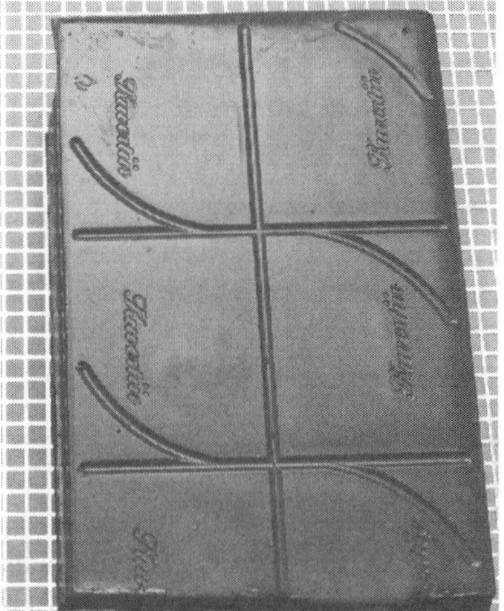

Exp 6.7
Tempered chocalate stored at constant temperature of 16°c.

Chocolate is best stored in an area of low humidity and stable temperatures of 15 to 18 °C (59–64.4°F) to avoid fat and sugar blooms (EXP 6.7).

Fat bloom and sugar bloom on the surface of chocolate products do not have any adverse health affect, but they make the products unappealing to consumers.

More Ideas to Try

Put 50 ml (1.6 oz) each of sunflower oil, olive oil, and peanut oil in separate bottles and keep the bottles in the refrigerator overnight. Observe the physical changes in oils that occur during storage. Discuss the results.

Study Questions

1. Explain the differences between saturated and unsaturated lipids.
2. Is coconut oil a fat or oil?

POINTS TO REMEMBER

- *Fats and oils are the organic compounds naturally obtained from animals and plants.*
- *Understanding the basic structure of fats and oils is crucial for chefs because the functional properties and shelf life of fats and oils are primarily related to their structures.*
-  *Fats and oils belong to a group of substances called lipids.*
- *The primary functional properties of lipids in food processing can be listed as:*
    a. Enhancement of flavor and mouthfeel
    b. Development of texture
    c. Shortening
    d. Emulsification
    e. Medium for transferring heat
    f. Development of appearance
- *Lipids with longer carbon chains have higher melting points.*
-  *The saturated fatty acids have higher melting points than unsaturated fatty acids of corresponding size.*
- *The saturated lipids have lower smoke points than unsaturated lipids.*
-  *Ideal frying lipids have a high smoke point that is well above the optimum frying temperature.*
-  *During storage, lipids and products rich in lipids may undergo lipid oxidation reaction.*
-  *The unsaturated lipids are more prone to lipid oxidation due to the double bonds in their structures.*
-  *Oxygen, heat, light, metal, and moisture increase the rate of oxidative rancidity.*
-  *The shortening power of the lipid determines the tenderness of the pastry.*
-  *A mixture of two or more liquids that are normally immiscible is called an emulsion.*
-  *Emulsions contain three components: a dispersed phase, a continuous phase, and the emulsifier.*
-  *Fat crystallization may develop on the surface of chocolate if it is not processed, stored, or served properly.*
-  *Tempering is a directed precrystallization process to promote the formation of the stable fat crystal forms in chocolates.*

SELECTED REFERENCES

Brown, A. C. 2007. *Understanding food: Principles and preparation*, 3rd ed. Belmont, CA: Wadsworth Publishing.

Costa, R. M., F. A. R. Oliveira, O. Delaney, and V. Gekas. 1999. Analysis of the heat transfer coefficient during potato frying. *Journal of Food Engineering* 39(3):293–299.

Costa, R. M., F. A. R. Oliveira, and G. Boutcheva, G. 2001. Structural changes and shrinkage of potato during frying. *International Journal of Food Science and Technology* 36:11–23.

Culinary Institute of America. 1995. *The new professional chef*, 6th ed. New York: John Wiley & Sons.

Functional properties of fats and oils. 2009. *Cereal Chemistry Journal* (AACC Int'l) 86(3):May/June. Online at: http://cerealchemistry.aaccnet.org/doi/pdf/10.1094/9780913250907.001

Ghotra, B. S., S. D. Dyal, and S. S. Narine. 2002. Lipid shortenings: A review. *Food Research International* 35(10):1015–1048.

Lonchampt, P., and R. W. Hartel. 2006. Surface bloom on improperly tempered chocolate. *European Journal of Lipid Science and Technology* 108(2):159–168.

McGee, H. 2004. *On food and cooking. The science and lore of the kitchen,* 1st revised ed.

Meulenaer B. D., and J. V. Camp. *Factors that affect fat uptake during French fries production*. Online at: http://www.euppa.eu/_files/factors-french-fries.pdf.

Ozilgen, S., and M. Ozilgen. 1990. Kinetic model of lipid oxidation in foods. *Journal of Food Science* 55:498–502.

Paul, S. R. 2006. Food frying. In *Encyclopedia of life support systems (EOLSS), food engineering theme*. Online at: http://www.eolss.net/Sample-Chapters/C10/E5-10-04-06.pdf.

University of California/Davis. *What is the smoke point of butter?* Online at: http://drinc.ucdavis.edu/dairychem.7htm.

Vaclavic, V. A., and E. W. Christian. 2008. *Essentials in food science,* 3rd ed. Berlin: Springer.

Weaver, C., and J. Daniel. 2003. *The food chemistry laboratory,* 2nd ed. Boca Raton, FL: CRC Press.

CHAPTER 7

Sensory Properties of Foods: Keys to Developing the Perfect Bite

REASONS TO DEVELOP A NEW FOOD PRODUCT

Competition in the food industry grows **tremendously** as a result of:

- Fast changing consumer trends.
- Increased consumer demand for new tastes.
- Increased consumer demand for safe food products.
- Increased consumer awareness of health.
- Increased consumer demand for specific food products, i.e., gluten-free food products, diabetic food products.
- Increased capacity of the production units that leads to producing more food. Companies can produce more varieties of similar food products in the same production area. For example, dairy producers can produce different types of cheeses at the same production area with the help of new processes and new equipment.

New food products and new processes need to be developed continuously to keep the company competitive in a changing food market. Food providers must work hard through innovations and improved processes to meet the consumer's expectations.

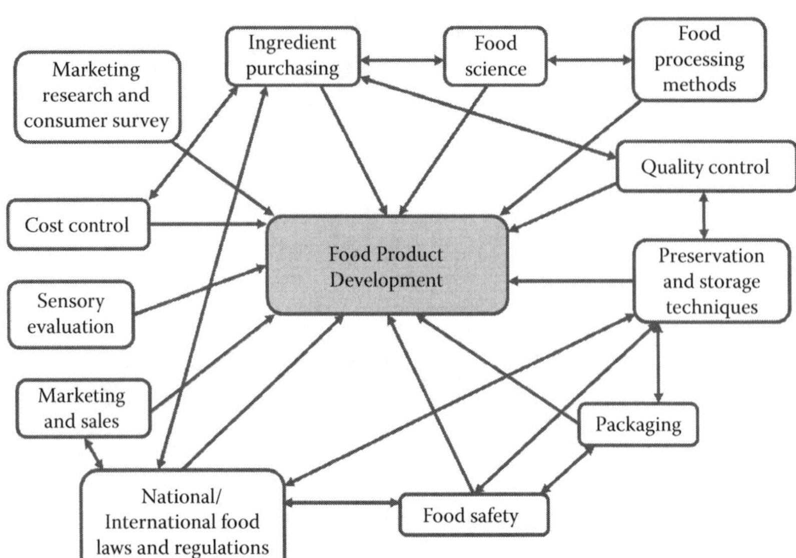

Figure 7.1
Product development is multidisciplinary team work and it requires knowledge of all items shown in the figure.

Companies invest money, staff, and equipment in new product development. Developing new food products is a *reward* for the company when it meets the consumers' needs and demands. On the other hand, possibility of failures of the new food products in the marketplace is a frightening *risk* for the companies. To minimize the risks, the companies should adopt a multidisciplinary approach. The new food product development process requires *multidisciplinary teamwork* (Figure 7.1).

In properly formed and managed product formulation teams, the individuals come from different disciplines, their contributions are properly harmonized, and they all work toward a common goal.

STAGES IN NEW FOOD PRODUCT DEVELOPMENT

The new food product development process may involve:

1. Formulation of completely new food products.
2. Modification of an existing food product, which is called *line extension*.

The same processing steps are followed in both cases, but line extension requires less financial resources and less time to develop a new food product.

The new food product development process has three major stages:

1. Idea development
2. Product development
3. Commercialization

Idea Development

The objective of this stage is to gather as many new ideas as possible. For the success of this stage:

- The mission of the organization must be considered.
- The goal of the company must be clearly identified.
- The consumer demands for new food products must be considered.
- Gaps in the market must be identified.

The team members carry out financial and technical reviews, feasibility studies, and legal analysis at this stage. The ideas that are weak in terms of their chances of market success should be filtered out at this stage. The most feasible ideas are carried to the product development stage.

Product Development

At this stage:

1. The expected quality attributes of the food product are identified.
2. The product is formulated. Specifications and the amounts of the ingredients are determined.
3. The processing steps and parameters are set.
4. A prototype is prepared.
5. Sensory evaluation tests are carried out.
6. If required, changes are carried out to the prototypes depending on the feedback from the sensory evaluation tests.
7. A trial production (pilot plant production) is carried out to produce large quantities of the improve prototype.
8. The prototypes from the pilot plant production are tested with a range of consumers through the sensory evaluation tests.
9. The product is either modified or discarded based on feedback gained from consumers through sensory evaluation tests. The information gained at this stage is used to move the production to the commercialization stage.

COMMERCIALIZATION

Once the consumers accept the product, the producers move forward to produce the product on a large scale. Finally, the product is introduced to the consumers in the marketplace (launching).

HOW TO CARRY OUT SENSORY EVALUATION TESTS

Sensory evaluation tests scientifically measure and analyze the consumers' responses to the sensory attributes of the food products, such as appearance, odor, sound, texture, and taste. They involve both qualitative observations and quantitative measurements that were previously explained in Chapter 1. In sensory evaluation tests, five senses are used to observe the characteristics of foods (PIC 7.1).

Sensory evaluation tests translate qualitative observations to quantitative data to

- improve the quality of food products;
- understand which attributes make the food products more acceptable to consumers;

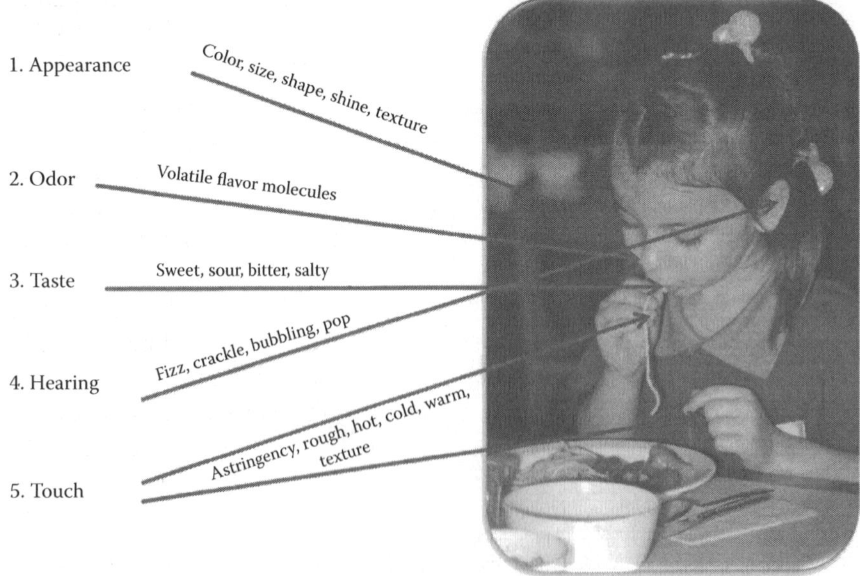

PIC 7.1

FIVE SENSES ARE USED TO OBSERVE THE CHARACTERISTICS OF FOODS: SIGHT, SMELL, TASTE, HEARING, AND TOUCH.

- understand whether differences exist between the ranges of food products;
- understand how big the differences are in a range of food products; and
- understand whether a final food product meets its targeted specifications.

Requirements for sensory evaluation tests can be listed as:

1. *Test environment*: Sensory evaluation tests should be carried out in a quiet area with adequate lighting and ventilation. Separate sections/booths that are free from distractions must be used for each panelist. The walls should be white or off-white in color.
2. *Timing*: Morning is the best time for sensory evaluation tests.
3. *Panelists*: Panelists are used to evaluate the food samples. The number and specifications of the panelists depend on the purpose of the study. Untrained (consumers), semitrained (experienced), or trained (highly experienced) panelists are incorporated in the tests depending on the purpose of the study. Willingness to participate, availability, food allergies, and food intolerances are the major factors to be considered when choosing the panelists. The panelists should be clearly informed about the purpose and the procedure of the sensory test. Potential consumers are the best panelists.
4. *Samples*: All food samples of the same type should be served to the panelists at the same temperature. The size, shape, and types of plates and/or cups should be the same for all samples of the same type. Size or volume served should be equal for all samples. The optimum number of samples to be tested in each session should be decided depending on the properties of the food. The samples must be coded to make sure that the panelists cannot recognize or identify the samples. Random three-digit alphanumerical codes, such as GP4, 534, 1K6, are usually used to label the samples. To avoid bias among the panelists, samples should not be coded A and Z or 1 and 10, since panelists might feel that samples A and 10 are better than samples Z and 1. Samples should be served in a random order; therefore, not all panelists will get the same samples at the same time. For example, if two samples are served, half of the panelists receive one sample first and the rest receive the other sample first.

Terminology is very important in sensory evolution tests. It must be consistent from product to product to minimize the panelist confusion and variability in defining the product attributes. Basic terms that are used in qualitative observations are given in Chapter 1.

Statistical techniques are often applied to the interpretation of the sensory test results. Basic statistical methods are given in Chapter 1.

Sensory evaluation tests are grouped into three categories based on the questions that they are addressing:

1. Discrimination/difference tests
2. Descriptive tests
3. Consumer acceptance/preference tests

Discrimination/difference tests analyze differences between food products. These tests are carried out to understand whether differences exist between the food products and/or how the consumers would define the difference between the products. For these tests, 25 to 50 untrained panelists are required.

The most common discrimination/difference tests include:

1. Paired comparison test
2. Triangle test
3. Duo-trio test

The ***paired comparison test*** is applied to compare two samples. Two coded samples are presented to the panelists. Panelists are asked to identify the sample that has a greater degree of intensity in terms of a specific sensorial attribute. Test scorecards are distributed to the panelists and they are asked to follow directions given on the scorecards.

For example, panelists are asked to determine which of two samples of Turkish Delights are sweeter.

PAIRED COMPARISION TEST SCORECARD	
You are provided with two samples. Taste the samples in the order from left to right and circle the sample that is sweeter.	
(Sample A)	(Sample B)
918	871

The ***triangle test*** is applied to understand if the differences between two samples are detectable. Three coded food samples are prepared, two of which are the same and the other is different. Samples are arranged in a triangle on a tray and presented to the panelists. Test scorecards are distributed to the panelists and they are asked to follow directions given on the scorecards.

For example, panelists are asked to choose the soup sample that is the most different from the other soup samples.

TRIANGLE TEST SCORECARD		
You are provided with three samples. Two of the samples are identical. Taste the samples in the order from left to right and circle the sample that is different than the order two samples.		
(Sample A)	(Sample B)	(Sample A)
871	63	918

The ***duo-trio test*** determines whether or not a sensory difference exists between two samples. This method is particularly useful when changes are done in ingredients, processing, packaging, or storage of food products currently available. Three food samples are prepared, two of which are the same. One of the two identical samples is marked as reference, and the other two samples are coded with alphanumeric codes. The samples are presented to the panelists. Test scorecards are distributed to the panelists and they are asked to follow directions given on the scorecards.

For example, panelists are asked to determine if there is a difference between an original apple pie recipe and the modified apple pie recipe.

DUO-TRIO TEST SCORECARD		
You are provided with three samples and one is marked as reference. One of the other samples is the same as the reference. Taste the samples in the order from left to right and circle the sample that is the same as the reference.		
(Reference, B)	(Sample A)	(Sample B)
	191	438

Descriptive tests are applied to detect how food products differ in preselected sensory attributes. For these tests, 8 to 12 trained panelists are required. Trained panelists are able to evaluate each attribute and the range of intensity of similar products. For example, a trained panelist presented a sample of chocolate would be able to rate the level of cocoa, cocoa butter, and the crystal structure similar to any instrument that would give a reading.

The most common descriptive tests include:

1. Descriptive rating test
2. Descriptive ranking test

The ***descriptive rating test*** is used to rate the intensity of preselected attributes of the food products. Test scorecards are distributed to the panelists and they are asked to follow directions given on the scorecards. The attributes of the food samples are generally rated on line scales or spider/star graphs.

For example, panelists are asked to rate the intensity of four attributes: crunchiness, texture, color, and taste of the chocolate bar.

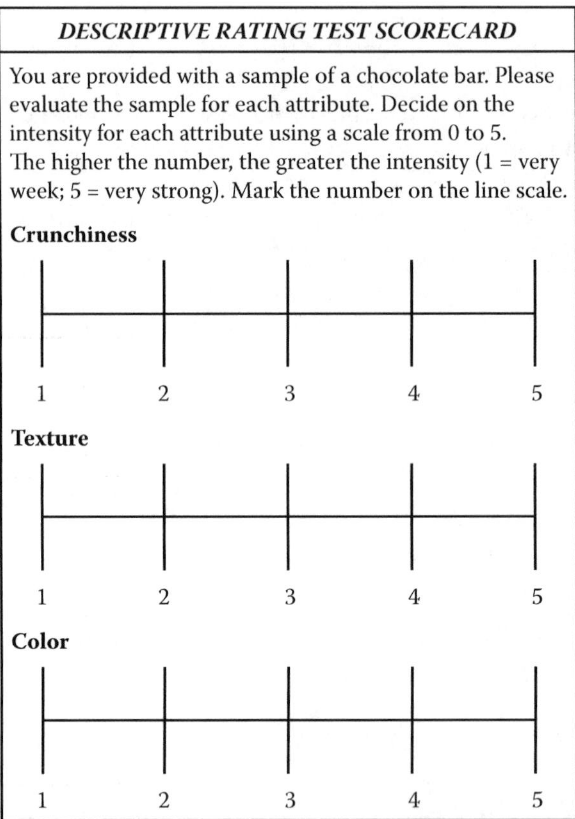

DESCRIPTIVE RATING TEST SCORECARD

You are provided with a sample of a chocolate bar. Please evaluate the sample for each attribute. Decide on the intensity for each attribute using a scale from 0 to 5. The higher the number, the greater the intensity (1 = very week; 5 = very strong). Mark the number on the line scale.

Crunchiness

1 2 3 4 5

Texture

1 2 3 4 5

Color

1 2 3 4 5

DESCRIPTIVE RATING TEST SCORECARD FOR TWO OR MORE SAMPLES

You are provided with two coded chocolate bar samples. Please evaluate the samples for each attribute. Decide on the intensity for each attribute using a scale from 0 to 10. The higher the number, the greater the intensity (1 = very week; 10 = very strong). Begin with one sample and evaluate it for all attribute. Taste the other sample for the same attributes.

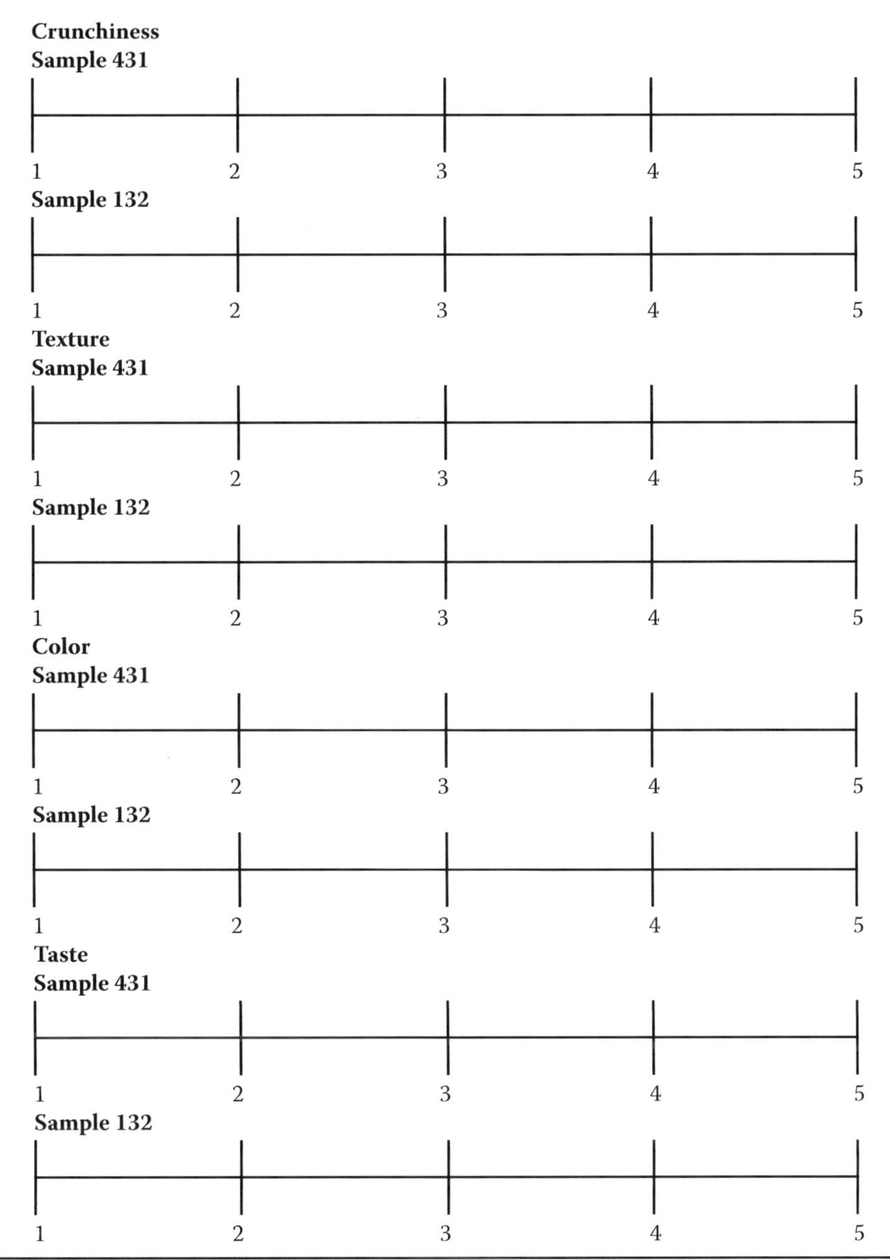

Crunchiness
Sample 431

1 2 3 4 5

Sample 132

1 2 3 4 5

Texture
Sample 431

1 2 3 4 5

Sample 132

1 2 3 4 5

Color
Sample 431

1 2 3 4 5

Sample 132

1 2 3 4 5

Taste
Sample 431

1 2 3 4 5

Sample 132

1 2 3 4 5

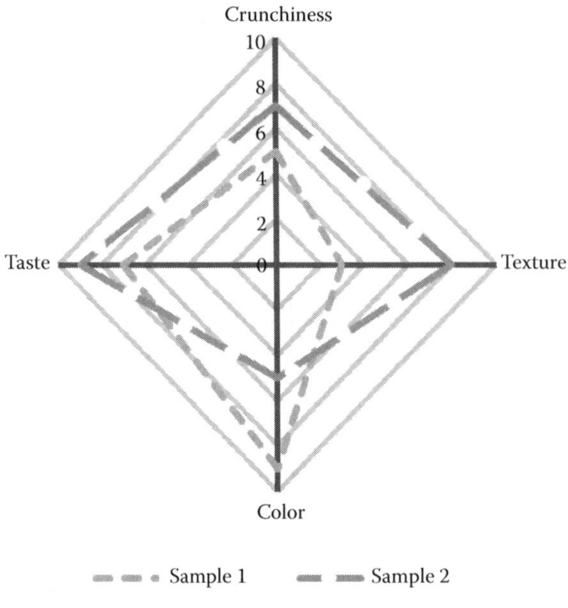

FIGURE 7.2
EXAMPLE OF A SPIDER GRAPH CHART.

Hint: To draw the spider/star chart, the average score of each attribute is calculated from the data obtained from each panelist and marked on the matching scale of the spider/star diagram. The marks are connected to draw the graph. The average scores of different products can be marked on the same chart to compare the sensory properties of the products. Shown in Figure 7.2 is an example of a spider graph.

The ***descriptive ranking test*** is applied to rank foods in the order of intensity of a specified attribute. Coded samples are presented to the panelists. Test scorecards are distributed to the panelists and they are asked to follow directions given on scorecards.

For example, panelists are asked to rank milkshake samples depending upon the intensity of sweetness.

RANK ORDER TEST SCORECARD

You are provided with five coded samples. Taste the samples and place them into rank order depending upon the intensity of sweetness. Write the code of the sample on the rank scale given below.

Not sweet	Slightly sweet	Moderately sweet	Very sweet	Extremely sweet

Consumer acceptance/preference tests are used to determine if the product is liked or disliked by the consumer. These tests also are used to determine the most preferred food product among the sets of similar products. For these tests, 25 to 50 untrained panelists (consumers) are required. Statistics are used to analyze the data.

The most common consumer acceptance tests include:

1. Preference tests
   a. Paired-preference
   b. Ranking for preference
2. Hedonic rating scale

The paired-preference test determines the preference of the consumer between two products. Two coded samples are presented to the panelists. Test scorecards are distributed to the panelists and they are asked to follow directions given on the scorecards.

For example, two lemon cake samples are presented to the panelists and they are asked to choose the one that they prefer.

PAIRED PREFERENCE TEST SCORECARD

You are provided with two coded samples. Taste the samples and circle the sample that you prefer.

(Sample A)	(Sample B)
345	762

The *ranking for preference test* ranks the food samples with respect to the preference degree. A number of coded samples are presented to the panelists. Test scorecards are distributed to the panelists and they are asked to follow directions given on the scorecards.

For example, panelists are asked to rank the five coded ice cream samples in order of their preferences.

RANKING FOR PREFERENCE TEST SCORECARD

You are provided with five coded samples. Taste the samples and rank them in an order of your preference. Please indicate the code your choices in the boxes given below.

	Sample code
1st choice	
2nd choice	
3rd choice	
4th choice	
5th choice	

The hedonic rating scale test is applied to measure the level of liking or disliking food products by the consumers. Coded samples are presented to the panelists. Test scorecards are distributed to the panelists and they are asked to follow directions given on scorecards.

For example, panelists are asked to indicate the extent of their liking for each of five meatball samples.

HEDONIC RANKING TEST SCORECARD

You are provided with five coded samples. Taste the samples and indicate your degree of liking. Check a box, from 1 to 9, to indicate your preference.

Sample code	1 dislike extremely	2 dislike very much	3 dislike moderately	4 dislike slightly	5 neither like or dislike	6 like slightly	7 like moderately	8 like very much	9 like extremely

POINTS TO REMEMBER

 *New products and new processes need to be developed continuously to keep the company competitive in a changing food market.*

 *A new food product development process requires a multidisciplinary teamwork.*

 *A new food product development process has three major stages:*

a. Idea development
b. Product development
c. Commercialization

 *Sensory evaluation tests scientifically measure and analyze the consumers' responses to the sensory attributes of the food products, such as appearance, odor, sound, texture, and taste.*

 *Sensory evaluation tests are grouped into three categories based on the questions that they are addressing:*

a. Discrimination/difference tests
b. Descriptive tests
c. Consumer acceptance/preference tests

 *Sensory evaluation tests provide excellent information for chefs to evaluate and improve their new dishes.*

SELECTED REFERENCES

British Nutrition Foundation. 2010. *Sensory evaluation: Teachers' guide*. Online at: http://www.foodafactoflife.org.uk/attachments/276dbf05-695c-44942bb55825.pdf

Crawford, I. M. 1997. *New product development*. In Marketing and Agribusiness texts. Rome: FAO (Food and Agriculture Organization). Online at: http://www.fao.org/docrep/004/w3240e/w3240e04.htm

Experimental design and sensory analysis. Pullman, WA: Washington State University. Online at: http://public.wsu.edu/~rasco/fshn4202005/EDSA.pdf.

Lawless, H. T., and H. Heymann. 2010. *Sensory evaluation of food: Principles and practices,* 2nd ed. Berlin: Springer.

Mason, R. M., and S. L. Nottingham. 2002. Sensory Evaluation Manual. *Alimentos Food*. online at: http://www.scribd.com/doc/890001/sensory

Naes, T., P. B. Brockhof, and O. Tomic. 2010. *Statistics for sensory and consumer science*. Hoboken, NJ: John Wiley & Sons.

O'Mahony, M. 1986. *Sensory evaluation of food: Statistical methods and procedures*. Boca Raton, FL: CRC Press.

Ozer, M. 2009. What do we know about new product idea selection? NC State University, Center for Innovation Management Studies. Online at: http://cims.ncsu.edu/downloads/Research/69_Ozer-%20New%20Prod.%20Idea%20Selection.pdf.

Ozilgen, S. 2011. Factors affecting taste perception and food choice. In *The sense of taste*, ed. E. J. Lynch and A. P. Petrov (pp. 115–126). Hauppauge, NY: Nova Science Publishers.

Rudolph, M. J. 2000. The food product development process. In *Developing new food products for a changing marketplace*. Boca Raton, FL: CRC Press.

Woods, T. W., and A. Demiralay. 1998. *An examination of new food product development processes: A comparative case study of two hazelnut candy manufacturers*. Lexington, KY: University of Kentucky. Online at: http://ageconsearch.umn.edu/bitstream/31979/1/sp980384.pdf.

CHAPTER 8

Food Additives in Culinary Transformations

CLASSIFICATION OF FOOD ADDITIVES

Food additives are any substances usually added to foods during production, preparation, processing, packaging, and transportation.

They become a part of the food product.

Food additives are generally divided into two groups: intentional/direct food additives and unintentional/indirect food additives.

INTENTIONAL/DIRECT FOOD ADDITIVES

They are often added during processing to perform a specific purpose. Intentional food additives are usually added to foods to:

1. *Maintain and improve nutritional quality.* Certain food additives are used either to replace vitamins and minerals lost in processing (enrichment) or to add nutrients to foods that may not have initially contained that nutrient (fortification). Fruit juices with vitamin C added are the best examples of enriched foods. Margarine is a good example of fortified foods because vitamin A, which is not naturally found in margarine, is added to it during processing.

2. *Preserve or improve quality and freshness.* Preservatives are used to extend the shelf life of food products. They decrease the rate of microbial growth and chemical reactions in foods. Organic acids, nitrites, sulfides, sugar, and salts are the best-known examples of antimicrobial preservatives. Antioxidants, such as vitamin C, BHA, and vitamin E maintain the color and flavor of the food products. They prevent rancidity in lipids and the products that are rich in lipids, such as cooking oils, cookies, and nut spreads. In addition, antioxidants keep peeled or cut fresh fruits and fresh vegetables from turning brown when they are exposed to air. The rate of spoilage and quality loss is usually higher than the rate expected by consumers when these additives are not added to foods during production.
3. *Help in processing or preparation.* Anticaking agents, humectants, leavening agents, maturing and bleaching agents, pH control agents, thickeners, and stabilizers are the most common food additives used to assist in processing and preparation of foods. For example, emulsifiers, such as lecithin, keep products from separating. Stabilizers and thickeners provide an even consistency or texture. Anticaking agents, such as silicone dioxide, keep foods from absorbing moisture. Humectants, such as glycerol, keep food moist and soft.
4. *Make food more appealing.* Some additives are added to improve, maintain, or enhance the taste, color, and aroma of foods. They are used in foods to replace the color and aroma lost during processing and storage; to produce a uniform product from raw materials that vary in color intensity; and to enhance the natural colors and aroma of foods.

Intentional/direct food additives may be natural, synthetic, or nature identical.

Natural food additives are primarily derived from plants and animals. For example, strawberries and beet roots are used to produce red food colorants, and carrots and turmeric are used to produce orange food colorants. Synthetic food additives do not occur in nature, but are made in factories. Petroleum-based chemical compounds are the basic sources of synthetic food additives. Nature-identical additives are factory-made copies of substances that occur naturally.

Unintentional/Indirect Food Additives

These are substances that enter foods in trace quantities during their production, processing, storage, or packaging. Indirect additives are commonly known as contaminants.

Unintentional/indirect food additives may enter foods through:

1. Polluted air, water, and soil, such as heavy metals.
2. Intentional use of various chemicals, such as pesticides, animal drugs, and fertilizers.
3. Natural sources, such as mycotoxins produced by certain types of microorganisms in foods.
4. A process, such as the chemicals that are formed during deep frying as a result of decomposition of lipids.

Indirect/unintentional food additives may imply a short-term or long-term risk to human health, and measures must be taken to minimize contaminants in foodstuffs. The summary of food additives is given in Table 8.1.

Consumers should feel confident about the safety of the foods that they consume. The intentional food additives must be considered safe in order to be used in food products. The safety of food additives is approved only on the basis of scientific studies and strictly regulated and controlled by governmental bodies. The risks and benefits of all food additives are scientifically assessed by considering the typical amount of consumption, chemical structure of the food additive, and short- and long-term health effects among different consumer groups, such as children, the elderly, and people with chronic diseases. All food additives are subject to ongoing safety studies. If new scientific studies suggest that a food additive is approved to be safe, but still may not be safe, the legal authorities may prohibit its use.

Food legislation in many countries require clear labeling of food additives in the list of ingredients. The E-code given to each food additive indicates that it is a "European Union approved" food additive.

Food additives cannot be used

1. to mask faulty processing and food spoilage;
2. if they decrease the nutritional value of the food products; or
3. if the same quality of the foods can be produced without food additives.

TABLE 8.1
Summary of Food Additives

Intentional/Direct Food Additives			
Purpose of Use	The Most Common Categories	Examples	Examples of Uses in Foods
To Maintain and Improve Nutritional Quality	Nutrients, primarily vitamins, minerals, essential fatty acids	Vitamin C, vitamin D, vitamin A, omega 3, iodine, calcium	Flour, breads, cereals, rice, macaroni, margarine, milk, fruit juices
To Preserve or Improve Quality and Freshness	Antioxidants	Citric acid, phosphoric acid, ascorbic acid, vitamin E, BHA, BHT	Oils and margarines, cereals
	Antimicrobial preservatives	Sorbates, sodium nitrite, calcium propionate, sulfites, salt, and sugar	Cured meats, sauces, baked goods, cheese, snack foods.
To Help in Processing or Preparation	Emulsifiers	Monoglycerides, diglyceride, lecithin, polysorbate 60	Salad dressings, mayonnaise, chocolate, margarine, butter
	Leavening agents	Sodium bicarbonate, calcium carbonate	Baked goods

	Stabilizers and thickeners	Starch, xanthan gum, agar, carrageenan, pectin, dextrins	Soups, sauces, ice cream, dairy products, puddings
	Humectants	Glycerin, sorbitol	Confections, shredded coconut
	Anticaking agents	Calcium silicate, silicon dioxide	Powdered sugar, salt
	Dough strengtheners and conditioners	Ammonium sulfate, L-cysteine	Baked goods
	Sweeteners and sugar replacers	Sorbitol, corn syrup, saccharin, aspartame, lactose, sucralose	Fruit juices, desserts, dairy products, many processed foods
	Fat replacers	Olestra, cellulose gels, carrageenan, microparticulated egg white protein, whey protein concentrate	Baked goods, dairy products, cakes, many reduced fat processed foods

Continued

TABLE 8.1 *(Continued)*
Summary of Food Additives

Intentional/Direct Food Additives			
Purpose of Use	The Most Common Categories	Examples	Examples of Uses in Foods
To Make Food More Appealing	Color	Blue 1 and 2, citrus red 2, yellow 5 and 6, paprika oleoresin, caramel color, saffron, fruit and vegetable juices	Confection, fruit juices, cakes, frostings, jams, many processed foods
	Flavoring	Quinine, spices, plant extracts	Snack foods, sauces, soda, confection, gelatin desserts, jams, marmalades
	Flavor enhancers	Monosodium glutamate, hydrolyzed vegetable protein	Dry soups, cereals, many processed foods
Indirect Additives			
The most common indirect food additives are the antibiotics, growth promoting hormones, pesticide residues, radioactive residues, chemicals from environmental pollution, polluted water, packaging materials, metals, and industrial wastes.			

EXPERIMENT 8.1

OBJECTIVE

To evaluate how color additives influence the acceptance of food products by consumers.

Lemonade Recipe

Ingredients and Equipment

- 400 g (14.1 oz) sugar
- 800ml Water
- 8–10 lemons
- Lemon rinds
- Yellow food coloring
- 2 pitchers
- 200 transparent drinking cups
- Saucepan
- Bowl
- Stove

Method

1. Squeeze juice from lemons into a bowl.
2. Heat the sugar, lemon rinds, and 500 ml (16.9 oz) of water in a saucepan until the sugar is dissolved completely. Set aside.
3. Discard rinds.
4. Combine the lemon juice, the sugar syrup, and rest of the water, and mix well.
5. Divide the lemonade into two pitchers.
6. Add two to three drops of yellow food coloring in one of the pitchers and mix well.
7. Refrigerate both pitchers until well chilled (both should come to the same temperature).
8. Carry out sensory analysis tests.

Sensory Test

Method

Prepare two different scorecards 8.1 and 8.2 for each of 60 consumers. Give codes to your sample i.e., 397 and 298. Carry out the acceptance sensory analysis test (follow the procedure explained in Chapter 7).

SCORECARD 8.1

You are Provided with Two Coded Samples. Taste the Samples and Indicate Your Degree of Liking. Check a Box, from 1 to 9, to Indicate Your Preference.

Sample Code	1 (Dislike extremely)	2 (Dislike very much)	3 (Dislike moderately)	4 (Dislike slightly)	5 (Neither like or dislike)	6 (Like slightly)	7 (Like moderately)	8 (Like very much)	9 (Like extremely)
397									
298									

SCORECARD 8.2

Check the Corresponding Boxes to Indicate the Reasons for Liking or Disliking Each Sample.

	Sample Number	
Attributes	397	298
It looks appealing		
It smells good		
It tastes great		
It is a good-quality product		
It does not look appealing		
It does not smell good		
It does not taste nice		
It is not a good-quality product		

Analysis of the Results

1. Calculate the average scores of scorecard 1 for each sample using the equation given in the first chapter.
2. Count the number of responses for each attribute of both samples and record in Data Table 8.1.
3. Draw a graph to compare the results for Data Table 8.1.
4. Compare and discuss the results.

DATA TABLE 8.1

Attributes	Total Number of Responses for Each Attribute	
	Sample 397	Sample 298
It looks appealing		
It smells good		
It tastes great		
It is a good-quality product		
It does not look appealing		
It does not smell good		
It does not taste nice		
It is not a good-quality product		

Tips: Let's assume the **overall responses** are found as given in Sample Table 8.1.

Figure 8.1 is how a **chart** on that dataset **would look.**

SAMPLE TABLE 8.1

Attributes	Total Number of Responses for Each Sample	
	Sample 397	Sample 298
It looks appealing	60	39
It smells good	55	43
It tastes great	52	24
It is a good-quality product	80	19
It does not look appealing	30	51
It does not smell good	25	47
It does not taste nice	38	66
It is not a good-quality product	10	71

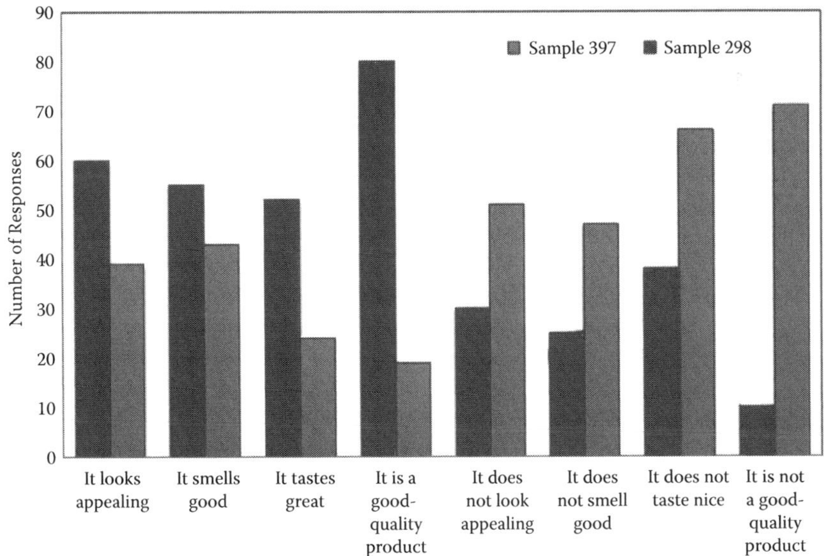

FIGURE 8.1

SAMPLE FIGURE TO COMPARE THE ATTRIBUTES OF THE SAMPLES.

More Ideas to Try

Repeat Experiment 8.1, but color the lemonade with blue food coloring instead of yellow. Analyze the consumer's attitudes toward the samples that have unusual colors. Compare your results with your results from Experiment 8.1.

POINTS TO REMEMBER

- *Food additives are any substances added to foods during production, preparation, processing, packaging, and transportation.*
- *Food additives are divided into two groups: intentional/direct and unintentional/indirect.*
- *Intentional/direct food additives are often added during processing to perform a specific purpose:*
  a. To maintain and improve nutritional quality
  b. To preserve or improve quality and freshness
  c. To help in processing or preparation
  d. To make food more appealing
- *Intentional/direct food additives may be natural, synthetic, or nature identical.*
- *Food additives are approved as safe only on the basis of scientific studies and are strictly regulated and controlled by governmental bodies.*
- *Food additives cannot be used:*
  a. To mask faulty processing and food spoilage
  b. If they decrease the nutritional value of the food products
  c. If the same quality of the foods can be produced without food additives
- *The E-code given to the food additive indicates that it is a "European Union approved" food additive.*
- *Unintentional/indirect additives enter foods in trace quantities during their production, processing, storage, or packaging.*
- *Antibiotics, growth-promoting hormones, pesticide residues, radioactive residues and chemicals from the packaging materials, metals, and industrial wastes are the most common unintentional/indirect additives in foods.*
- *Unintentional/indirect additives may show short- or long-term adverse health effects.*

SELECTED REFERENCES

Brown, A. C. 2007. *Understanding food: Principles and preparation*, 3rd ed. Belmont, CA: Wadsworth Publishing.

International Food Information Council (IFIC) and U.S. Food and Drug Administration. 2004, revised 2010. *Food ingredients and colors*. Online at: www.fda.gov/downloads/Food/FoodIngredientsPackaging/ucm094249.pdf.

McGee, H. 2004. *On food and cooking. The science and lore of the kitchen,* 1st Scribner revised edition. New York: Scribner.

McGee, H. 2010. *Modern gastronomy A to Z*. Boca Raton, FL: CRC Press/Taylor & Francis Group.

Purdue University. 2002. *Food additives*. Online at: http://www.four_h.purdue.edu/foods/Food%20additives.htm.

Vaclavic, V.A., and E. W. Christian 2008. *Essentials in food science,* 3rd ed. Berlin: Springer.

Glossary

Amino acid: Monomers of proteins. They contain an amino (NH_2) group and a carboxyl (COOH) group. Amino acids are joined together by peptide bonds to form polypeptide chains.

Amorphous: Has no crystalline structure, such as lollipops, marshmallows, and cotton candies.

Amylase: An enzyme (protein) that breaks down starch.

Amylopectin: A fraction of starch. It is a highly branched polymer of glucose.

Amylose: A fraction of starch. It is a linear chain polymer of glucose.

Antioxidant: A substance that prevents or slows down oxidation reaction by interfering with the chain reaction.

Atom: The smallest particle of a substance having the chemical properties of the substance. An atom consists of a central nucleus surrounded by negatively charged electrons. The nucleus is positively charged, and contains positively charged protons and uncharged neutrons. The number of electrons, protons, and neutrons of each atom determines the chemical reactions that the atom may have with other atoms.

Blanching: A cooking technique in which food is briefly immersed in boiling water for a few seconds and then placed in cold water for sharp cooling. The primary purpose of blanching is to deactivate the enzymes in foods. It is usually applied to prepare fruits and vegetables for long-term freezer storage.

Boiling point: The temperature at which a compound changes from a liquid to a gas.

Bound water: Type of water, which cannot be easily removed from the food. It is not available for chemical and biological reactions.

Caramelization: A nonenzymatic browning reaction. It is a complex set of sugar degradation reactions, which happens only at high temperatures.

Casein: A milk protein.

Chemical change: An irreversible change. A chemical reaction occurs and a new substance is formed. Cooking eggs, for instance, is an example of a chemical change; the egg white and egg yolk change from liquid to solid.

Coagulation: Transformation of native protein structure into a soft semisolid or solid mass. Coagulation of protein occurs when the denaturation process reaches an advanced state. Once proteins are coagulated, they cannot be returned to their native state.

Compound: Made up of atoms of two or more different elements bound together, such as carbohydrates and fats. Components may be separated by chemical means. Atoms of elements combine in fixed ratios to form compounds.

Covalent bond: A strong chemical bond that joins two atoms together by the sharing of electrons in the outer atomic orbitals.

Crystallization: The formation of solid crystals from a solution in a precise orderly structure.

Denaturation: Change of protein structure from its natural state. Denaturation involves the disruption of the secondary, tertiary, and quaternary structures of proteins.

Dextrinization: The breakdown of starch polymers into smaller molecules when heated to high temperatures without liquid.

Disaccharide: A carbohydrate that is composed of two molecules of simple sugars (monosaccharide) linked to each other.

Element: One of the basic substances that are made of atoms of only one kind and that cannot be separated by ordinary chemical means into simpler substances.

Emulsion: A mixture of two immiscible liquids in which minute droplets of one liquid are dispersed in another.

Enzymatic browning: A chemical change. It occurs when the phenolic substances in foods react with oxygen, in the presence of specific types of enzymes. Browning of foods occurs as a result of enzymatic browning reaction, e.g., the browning of peeled apples.

Enzyme: A chemical substance produced by living organisms, such as animals, plants, and microorganisms. Enzymes are capable of initiating certain chemical changes and/or increasing the rate of chemical reactions.

Fats: Substances that are composed of glycerol and fatty acids. They are solid at room temperature. Animals are the primary sources of fats.

Fatty acids: Building blocks of fats and oils. They are carboxylic acids with long hydrocarbon chains. Fatty acids are either saturated or unsaturated.

Fermentation: Transformation of sugar primarily into acids, gases, and/or alcohol by the action of microorganisms, e.g., lactic acid bacteria ferment lactose to lactic acid in the yogurt manufacturing process.

Fibrous proteins: Proteins that are composed of polypeptide chains assembled along the straight axis, e.g., collagen and elastin. They are shaped like rods or wire and are usually insoluble. They are primarily found in meat, chicken, fish, and wheat.

Food additives: Substances that are directly or indirectly added to foods during food processing.

Free water: Available water for chemical and biological reactions. Not chemically or physically bound. Free water can be easily removed from food by cutting, squeezing, pressing, or drying.

Freezing point: The temperature at which a liquid changes to a solid.

Fructose: A simple monosaccharide that is known as a fruit sugar. It occurs naturally in many fruits.

Globular proteins: Proteins composed of polypeptide chains twisted into a rounded, compact shape as globs or spheres. They are usually soluble and are primarily found in milk and eggs.

Glucose: A simple monosaccharide (sugar).

Gluten: A stretchy cereal protein that gives structure to baked goods.

Glycerol: An alcohol.

Hydrocarbon: Substance made of hydrogen and carbon, and is hydrophobic.

Hydrogen bonding: The attractive interaction of a hydrogen atom bearing a partial positive charge, which is covalently bonded to an electronegative atom, with another electronegative atom, such as fluorine, nitrogen, or oxygen that comes from another molecule. It is weaker than an ionic bond or covalent bond.

Hydrogenation: A chemical process that adds hydrogen atoms to an unsaturated oil to saturate it. Liquid oils become solid fats. Examples of hydrogenated fats include margarine and vegetable shortening.

Hydrolysis: A chemical reaction that breaks down the bonds in molecules using water.

Hydrophilic: Substances that have an affinity for water, and are water-soluble.

Hydrophobic: Substances that do not dissolve easily in water.

Insoluble: A substance that is not capable of being dissolved; for example, fats are insoluble in water.

Ionic bonding: The attractive interaction between two oppositely charged (+ and –) ions, e.g., common table salt (sodium chloride). Ionic bonding involves the complete transfer of electron(s) between atoms that generates two oppositely charged ions. Oppositely charged ions are attracted to each other by electrostatic forces, thus an ionic bond is formed.

Lactose: A disaccharide, which is made from galactose and glucose units. It is known as a milk sugar.

Lard: A rendered fat obtained from swine.

Lecithin: An emulsifier found in eggs and soybean oil.

Lipids: Organic substances that are insoluble in water and soluble in nonpolar solvents. Fats and oils are known as lipids.

Maillard browning reaction: A nonenzymatic browning reaction caused by the reaction of reducing sugars with proteins and amino acids in foods. The Maillard reaction is responsible for the brown color and nutty flavor of cooked foods, e.g., roasted meat, toast, and coffee.

Melting point: The temperature at which solids becomes liquids.

Mixture: Made of two or more different substances that are physically mixed with each other. Mixtures can be separated into their components by physical means.

Molecule: Two or more atoms join together chemically to form molecules.

Monomer: A molecule that can be bonded to other identical molecules to form a polymer.

Monosaccharide: A simple sugar that constitutes the building blocks of disaccharides, oligosaccharides, and polysaccharides. Examples are glucose, fructose, and galactose.

Monounsaturated fatty acid: Fatty acids that contain one C = C double bond in the hydrocarbon chain.

Oil: Substances that are composed of glycerol and fatty acids. They are liquid at room temperature. Plants are the primary sources of oils.

Osmosis: The movement of water through a semipermeable membrane from a region of high solvent concentration to a region of lower solvent concentration.

Pectin: A large polysaccharide that is present in plant cells. It occurs naturally in most fruits, and is concentrated in the skins and pips of different types of fruits.

Peptide bonds: Bonds that connect the amino group of one amino acid and the carboxylic acid group of another amino acid.

Peptides: Short chains of amino acids.

Physical change: A reversible change that involves a change from one state (solid or liquid or gas) to another without a change in chemical composition, e.g., ice melting. No new substances are formed.

Polymers: Macromolecules that contain 10 or more monomer units. Protein is a polymer of the monomer amino acids.

Polypeptides: Long chains of amino acids.

Polysaccharide: A carbohydrate that is made up of a number of simple sugar units (monosaccharide) bonded together.

Polyunsaturated fatty acids: Fatty acids that have multiple C = C double bonds.

Primary structure of proteins: The sequential order of amino acids in a protein.

Proteins: Complex polymers composed of amino acid monomers.

Quaternary structure of proteins: Two or more polypeptide chains joined together to form the quaternary structure of proteins.

Rancidity: The chemical deterioration of lipids. Undesirable odors and flavor may develop as a result of chain reactions. Rancidity can be either enzymatic or oxidative.

Rennet: A combination of different proteolytic enzymes. It is usually only extracted from the stomachs of young animals.

Rennin: A proteolytic enzyme found in rennet. It curdles milk.

Retrogradation: Starch molecules, particularly the amylose fraction, reassociate in an ordered structure in gelatinized starch; eventually a crystalline order appears and water is squeezed out, e.g., staling of bread.

Saturated fat: A fat that is normally solid at room temperature, e.g., butter. Fatty acids in their structures are saturated; they contain only carbon–carbon (c–c) single bonds in the hydrocarbon chain.

Saturated solution: A solution in which the dissolved solute is in equilibrium with the undissolved solute.

Secondary structure of proteins: The three-dimensional organization of the polypeptide chain formed by intramolecular and intermolecular hydrogen bonding.

Smoke point: The temperature at which heated fats and oils begin to smoke and emit unpleasant odors.

Soluble: A substance that is capable of being dissolved, e.g., salt is soluble in water.

Solute: Substances in solutions.

Solution: A mixture of one or more solutes and a solvent.

Solvent: The substance in which a solute dissolves.

Starch: A polymer of glucose units. It has two fractions: amylose and amylopectin.

Sucrose: A disaccharide made from glucose and fructose units.

Tertiary structure of proteins: The three-dimensional organization of the polypeptide chain, which is maintained by weak, noncovalent interactions, such as hydrophobic interactions and salt bridges.

Unsaturated fat: A lipid that is normally liquid at room temperature, e.g., olive oil. Fatty acids in their structures are unsaturated; they contain C = C double bonds in the hydrocarbon chain. Those lipids are normally liquid at room temperature, e.g., olive oil.

Viscosity: The resistance to flow in a liquid.

Water activity: Refers to free water in foods.

Water holding capacity: Ability of a food structure to entrap water.

Whey: One of the two major proteins in milk. It is the liquid left behind after milk has been curdled and strained as in cheese making.

INDEX

L

M

N

O

T

U

V

W

Y